AF720153

Setting the Table

Florida Museum of Natural History: Ripley P. Bullen Series

FLORIDA MUSEUM
OF NATURAL HISTORY

SETTING THE TABLE

Ceramics, Dining, and Cultural Exchange in Andalucía and La Florida

KATHRYN L. NESS

University of Florida Press
Gainesville

Copyright 2017 by Kathryn L. Ness
All rights reserved
Printed in the United States of America on acid-free paper

This book may be available in an electronic edition.

22 21 20 19 18 17 6 5 4 3 2 1

Library of Congress Cataloging-in-Publication Data
Names: Ness, Kathryn L., author.
Title: Setting the table : ceramics, dining, and cultural exchange in Andalucía and La Florida / Kathryn L. Ness.
Other titles: Ripley P. Bullen series.
Description: Gainesville : University of Florida Press, 2017. | Series: Florida Museum of Natural History: Ripley P. Bullen series | Includes bibliographical references and index.
Identifiers: LCCN 2016038500 | ISBN 9781683400042 (cloth)
Subjects: LCSH: Saint Augustine (Fla.)—Social life and customs—18th century. | Andalusia (Spain)—Social life and customs—18th century. | Ethnoarchaeology—Florida—Saint Augustine. | Ethnoarchaeology—Spain—Andalusia. | Arts, Latin American—History. | Ceramics—Florida—Saint Augustine—History. | Ceramics—Spain—Andalusia—History. | Dinners and dining—Florida—Saint Augustine—History. | Dinners and dining—Spain—Andalusia—History. | Cultural relations.
Classification: LCC F319.S2 N47 2016 | DDC 946.8—dc23
LC record available at https://lccn.loc.gov/2016038500

UF PRESS
UNIVERSITY OF FLORIDA

University of Florida Press
15 Northwest 15th Street
Gainesville, FL 32611-2079
http://upress.ufl.edu

To Kevin

Contents

Figures

Tables

Acknowledgments

In the prologue of his immortal work *Don Quixote*, Miguel de Cervantes Saavedra wrote that he was about to give up the book when a friend came along and gave him some lengthy encouragement and humorous, albeit questionable, suggestions about finding sources. While I avoided his friend's advice about making up sources, I can sympathize with Cervantes's gratitude for such a champion. In writing this book, I have had many such supporters, and I am infinitely grateful for each and every one of them.

This project would not have been possible without the help of several organizations and individuals on both sides of the Atlantic. In Spain, I owe a tremendous thanks to the Junta de Andalucía, the Museo Arqueológico Municipal de Jerez de la Frontera, Rosalía González Rodríguez, Francisco Barrionuevo, Ester López Rosendo, and José Ruiz Gil.

In the United States, I have to thank Kathleen Deagan, Charles Cobb, Gifford Waters, Ann Cordell, Carl Halbirt, Mary Beaudry, Amalia Perez-Juez, Fred Smith, Andrea Berlin, Mac Marston, Russell Skowronek, Charles Ewan, James Cusick, Charles Tingley, the staff of the St. Augustine Historical Society Research Library, Lillian Lambrechts, Jaidee Easley, Ellie Jordan, and Kay Ueda for their support and guidance.

I am very grateful for generous financial support from the National Science Foundation (DDRIG BCS-1340232), the Boston University Humanities Foundation, the Cora du Bois Charitable Trust, the Department of Archaeology at Boston University, and the Writing Program at Boston University. I am also indebted to Charles E. Orser, who accepted an edited version of chapter 3 for publication in the June 2015 issue of the *International Journal of Historical Archaeology*. Likewise, I owe my sincere gratitude to those at the University of Florida Press who took an interest in this project and helped turn my research into this book.

Finally, I have to thank the many, many friends and family members who took an active interest in this project. Kevin, Jocelyn and David, Beverly and Tom, and Sandy and Jean deserve my deepest gratitude for their support and assistance. It is impossible to name everyone who played a role in this project, but they know who they are and, hopefully, know just how grateful I am for their support. Thank you all.

Abbreviations

AGI	Archivo General de Indias
COSA	Ceramic Organization of the Spanish Atlantic
FLMNH	Florida Museum of Natural History
SAAA	St. Augustine Archaeological Association

1

A Changing Spanish Identity

> It is no wonder that [. . .] Cervantes's hero [Don Quixote] has become a symbol of Spain and Spanish culture.
>
> Robert Bayliss, *What* Don Quixote *Means (Today)*, 387

In the eighteenth century, Spaniards and Spanish Americans found themselves embedded in changing political, economic, and cultural environments that forced them both to negotiate new ideas and goods and to define what it meant to be "Spanish." A Spanish individual living on either side of the Atlantic, for example, might have adopted French fashions and phrases in imitation of the new Bourbon court in Madrid, while enjoying American foods like turkey on British dishes that were meant to imitate Asian porcelain. While some individuals attempted to integrate these new influences into their lives and redefine their sense of cultural identity, others protested the adoption of non-Spanish influences by promoting customs that they perceived as characteristically Spanish.

This struggle to delineate Spanish cultural identity can perhaps best be seen in the iconic Spanish novel *Don Quixote de la Mancha* by Miguel Cervantes de Saavedra. During the eighteenth century, the book regained popularity after having been out of print from 1674 to 1704. Despite being translated into numerous languages and enjoyed throughout Europe in the century following its publication in 1605, it was not until 1780 that the Spanish Royal Academy accepted it as a classic piece of literature and printed a new edition. While some scholars have suggested that the lack of new editions in the late seventeenth century could be related to criticisms of the novel during this period (Santos 2012), its resurgence is also an example of the struggle to define Spanish identity and cultural heritage during this period.

Cervantes's novel, as humorous and fictitious as it is, is both quintessentially Spanish and reflective of the cultural influences that impacted early modern Spanish society. The knight's preoccupation with status and earning the appropriate social standing to be worthy of his love, the "peerless Dulcinea," for example, is indicative of contemporary Spanish attitudes toward social structure and class. The casual references to the Americas and non-Christians, in contrast, hint at the external cultural influences that had been present in the Iberian Peninsula for generations.

This book is not, however, an analysis of *Don Quixote* but rather a study that relies on the knight to escort us through the cultural negotiations of the early modern Spaniards and Spanish Americans who might have read and enjoyed the adventures of this unusual gentleman and his faithful squire, Sancho Panzo. On both sides of the Atlantic, these nonfictitious individuals attempted to amalgamate new ideas and fashions from the Americas, Asia, and other parts of Europe with their Spanish heritage and identity. In doing so, they created a distinct Spanish Atlantic identity that was neither wholly Spanish nor wholly Spanish American, but rather reflected the increasing globalization of the time while also maintaining some Spanish traditions.

Although Spaniards and Spanish Americans constituted a significant portion of the Atlantic world and represented one of the largest and most powerful early modern imperial enterprises, they have been largely underrepresented in trans-Atlantic studies. While numerous scholars have addressed British and British-American history and archaeology (Armitage and Braddick 2002), there are currently only two archaeological studies of the Spanish Atlantic (Lister and Lister 1987; McEwan 1989), neither of which addresses domestic sites or the eighteenth-century Spanish Empire.

One reason for this lack of comparative research is that, to date, the majority of scholarly work regarding early modern Spanish sites concentrates on settlements in the Americas, while Spanish archaeologists have only recently begun to investigate sites from this time period. This state of imbalanced research has resulted in several unfortunate consequences. Because scholarship in the Americas has been conducted without the aid of a comparable data set from Spain, interpretations of Spanish American sites have relied on conjectural ideas of contemporary Spanish lifeways, rather than on models established from empirical archaeological research. Additionally, differences in research methods and approaches to studying and interpreting the data have resulted in incompatible data sets between Spain and her former territories, which makes such trans-Atlantic research difficult at best.

Despite these challenges, the situation is improving. An increasing number of archaeologists in Spain are studying the *moderna* (1492–1809) and *contemporanea* (1809–present) eras and collaborating with American colleagues, while researchers in other fields, such as history and art history, are also conducting trans-Atlantic studies. Although these scholars focus on the Spanish Empire, the results of their research and their new methodological approaches for this type of work are beneficial to others studying the Atlantic world. The more we know about early modern Spaniards and Spanish Americans, the better equipped we will be to study their role in Atlantic history and how they contributed to, participated in, and shaped political, cultural, and economic policies throughout the region. This book, therefore, represents one possible way of exploring how individuals in Spain and the Americas lived in the eighteenth century and how the multidirectional nature of colonialism impacted their representations of cultural identity and heritage.

The Sites and Research

This study focuses on three eighteenth-century domestic sites in Spain and Florida. The first site, La Calle Corredera, is located in Jerez de la Frontera, Spain and was most likely occupied by a series of middle-class families during the eighteenth and nineteenth centuries. The second site, Palm Row, belonged to Chief Adjutant Francisco Ponce de León and his family, who lived in St. Augustine, Florida in the middle of the eighteenth century; they left the town in 1763 when the territory was ceded to the British. While the last site, known as the Monson Motor Lodge, was occupied by numerous individuals who lived in St. Augustine in the mid-1700s, this study uses the evidence for one phase of occupation that corresponds to Spanish-born merchant Juan de Salas and his family in the 1760s.

La Calle Corredera and the Ponce de León and the de Salas households are indicative of the kinds of individuals who populated the Spanish Atlantic and contributed to developing this Spanish Atlantic identity, either in person or through long-distance exchanges. While earlier Spaniards saw the Americas as a place to make their fortune before returning home to a life of leisure in Spain, the eighteenth-century families who lived at these sites saw the Americas as legitimate parts of the Spanish Empire. For the Ponce de León and the de Salas families, St. Augustine was very much a home rather than a temporary settlement until they could afford to return to Spain. The family at La Calle Corredera, in turn, decorated their house with Mexican ceramics and enjoyed American foodstuffs such as chocolate. Together, these individuals are prime examples of how Spaniards and Spanish Americans navigated the cultural developments of the time.

The present study focuses on the ceramics recovered at these sites because of their predominance in the archaeological record and their ability to serve as proxies for social activities such as entertaining, cooking, and dining. For modern scholars, these vessels are valuable sources of information, as they can provide clues to foodways, fashions, aesthetics, and cultural trends and, when compared among sites, can indicate cultural exchanges and behavioral differences.

These items, however, represented only a small percentage of the average household's goods. Indeed, in *moderna*-era Spanish paintings, ceramics seem rarely to be depicted, unless the image is of a still life or shows dining or cooking, such as in *casta* paintings. Similarly, references to ceramics are somewhat rare in texts of the period; they are hardly mentioned in *Don Quixote*, and household inventories inconsistently document ceramic objects. Because early modern consumers appear to have viewed ceramics as relatively disposable, somewhat unimportant elements of a household assemblage, it is also important to include other lines of evidence and consider both the items that went into and onto the ceramics as well as those that surrounded the vessels, in order to obtain a more holistic view of the past.

Consequently, this study also includes nonceramic elements, such as early modern texts, food, aesthetics, and ideas, all of which might have had an impact on the function, shape, use, or decoration of the vessels. By combining archaeological and documentary research, we can create a snapshot of both the individual sites in question and life in eighteenth-century Jerez de la Frontera and St. Augustine.

Based on this glimpse of life in the Spanish Atlantic, many early modern Spaniards and Spanish Americans were very willing to incorporate and adapt a variety of cultural influences. In Spain, individuals of various social ranks strove to copy the French fashions popular in Madrid.

This can be seen in the clothing, food, tableware, and even etiquette of the period. These trends quickly made their way across the Atlantic to places such as St. Augustine, where families like the Ponce de Leóns and de Salases used tableware that was similar to that used by their counterparts in Jerez and most likely also followed other vogues from Spain.

Spanish Americans were not, however, passive absorbers of European fashion. Successful Spanish Americans sometimes returned to Europe with their wealth and American customs, while others sent foods, ceramics, and raw materials to Spain. As early as 1568, Mexican furniture was being shipped to Spain and used by elite Spaniards such as the Duke of Medina Sidonia (Rivas Pérez 2013:59). Although the Americas might not feature prominently in early modern Spanish literature such as *Don Quixote*, they helped shape early modern Spanish culture in many ways.

While this amalgamation of cultural influences, which included goods and ideas from throughout the Spanish Atlantic, is reflected in many aspects of daily life, it is especially apparent in the dining practices and tableware of the period. New vessel forms, cooking styles, and ingredients became popular in the eighteenth century, while new ceramic technologies and industries helped spread new ideas and aesthetics.

Organization of This Book

Chapter 2 situates this study in Spanish Atlantic history and culture. As with any imperial situation, economic policies and cultural traditions impacted both the colony and the motherland. In the case of Spain, trade with the Americas resulted in economic prosperity in Andalucía and offered new opportunities for individuals who wished to emigrate to the Americas. These same trade regulations, however, proved problematic for Spanish Americans and affected the relationship between more remote parts of the Spanish Empire, such as St. Augustine, and the Spanish homeland. Despite their economic differences, individuals on both sides of the Atlantic viewed themselves as Spanish citizens and were acutely aware of their place in society, a position which was both socially and legally defined according to one's religious and biological heritage. This background helps set the stage for future chapters as it explores the long-lasting trans-Atlantic connections that endured into the mid-1700s, as well as the preoccupation with status and Spanish identity that pervaded Spanish Atlantic society during the early period of the Bourbon dynasty.

Chapter 3 outlines the study's methodological approach to studying Spanish ceramics by focusing on vessel form and purpose. One of the most significant challenges to conducting trans-Atlantic studies on Spanish sites is the lack of comparable data. Spanish archaeologists, for example, document vessel forms and general ceramic types, while American archaeologists record ceramic types according to a very specific, detailed typology yet sometimes neglect to consider vessel forms. As a result, one might be able to obtain information about sites on both sides of the Atlantic, but such data cannot be directly compared because of the radically different approaches to inventory and study of the artifacts. The ceramic classification system presented here, Ceramic Organization in the Spanish Atlantic (COSA), is designed to remedy this issue by providing one way to identify ceramics from Spanish and Spanish colonial sites and therefore facilitate both this and future studies. While a more detailed discussion of COSA has been published elsewhere (Ness 2015a), this chapter serves as an overview of the classification system to facilitate the discussions in later chapters.

Chapter 4 presents La Calle Corredera in Jerez de la Frontera. The town has a long history of connections to Spanish America, especially Florida. Álvar Núñez Cabeza de Vaca, one of the earliest Spanish explorers to visit and document his travels in Florida, was born in Jerez de la Frontera, as was at least one eighteenth-century resident of St. Augustine. These connections, as well as the town's economic boom in the early eighteenth century as a result of changes in Spanish American trade routes and regulations, meant that the town and its residents were deeply involved in and influenced by the Atlantic world and had strong ties to the Americas. La Calle Corredera, in turn, represents what a standard, middle-class household in this town might have owned and used. Its location in an eighteenth-century neighborhood and status as one of the few early modern sites to have been excavated and published in Andalucía make it ideal for this project.

Chapter 5 introduces the Ponce de León and de Salas households in St. Augustine. Since its inception in 1565, this military town on the northeast coast of Florida represented one edge of Spain's extensive empire and was intended to serve as a vanguard for Spanish interests in the region. While modern historians often describe St. Augustine as a backwater outpost that was frequently forgotten or neglected by officials in Spain and Mexico, it had strong connections to the mother country, with Spanish-born individuals immigrating to the town well into the eighteenth century. One such individual was Juan de Salas, whose house and store are discussed in this chapter. Evidence from his property and from the Ponce de León household provide further indications that, despite infrequent supply shipments and occasional shortages, the town's residents were well aware of popular fashions in Spain and largely able to obtain the items necessary to follow the latest trends.

Chapter 6 explores the evidence of sociocultural change on both sides of the Atlantic, specifically as indicated by ceramics from these sites. While utilitarian cooking vessels in Spain appear to have remained largely the same between the seventeenth and eighteenth centuries, new tableware forms suggest that the food being cooked in these vessels changed dramatically during the 1700s. Documentary evidence indicates that these shifting foodways were influenced by an increased use of American foods, as well as a rise in French cooking methods. In the face of these widespread changes, however, some more traditional Spanish elements did remain the same, including vessels related to hygiene, which appear to have maintained the same form from the seventeenth through the nineteenth centuries. The blend of change and continuity indicated that Spaniards and Spanish Americans were selectively adopting new objects, ideas, and habits and combining them with familiar elements, thereby refining what it meant to be both Spanish and part of the early modern world.

Chapter 7 contextualizes these findings in the larger world of Spanish and Spanish American archaeology. The similarities in vessel forms and the nearly concurrent changes on both sides of the Atlantic indicate that Spanish Americans continued to look to Spain as a center of culture and fashion, even as late as the 1760s. The evidence for such close connections between Spain and Spanish Florida forces us to refine our interpretations of Spanish St. Augustine and reevaluate how the residents expressed their heritage and viewed their status as citizens of the Spanish Empire. Similarly, the evidence for changes in dining practices contradicts contemporary and some modern discussions of eighteenth-century Spanish foodways and indicates that perhaps non-Spanish influences had a more widespread impact than previously thought.

2

Spanish Atlantic History and Culture

> Porque te hago saber, Sancho, que hay dos maneras de linajes en el mundo: unos que traen y derriban su decendencia de príncipes y monarcas, a quien poco a poco el tiempo ha deshecho, y han acabado en punta, como pirámide puesta al revés; otros tuvieron principio de gente baja, y van subiendo de grado en grado, hasta llegar a ser grandes señores. De manera que está la diferencia en que unos fueron, que ya no son, y otros son, que ya no fueron . . .
>
> [Because you know, Sancho, that there are two kinds of lineages in the world: some who trace their descent from princes and kings, who little by little over time have been exhausted and have finished in a point, like an upside-down pyramid; others who began from lowly people and are gradually raising themselves until they get to be great gentlemen. The difference is that some were but are no longer, and others are but were not . . .] (Author's translation)
>
> Miguel de Cervantes Saavedra, *Don Quixote*, Volume I, Chapter XXI

Heritage and family background were of paramount importance in early modern Spain. On an individual level, they could dramatically impact one's economic and social prospects. On a social level, there was a strong preoccupation with defining what it meant to be "Spanish," especially during the eighteenth century, when Spaniards on both sides of the Atlantic faced a plethora of new outside cultural influences from a variety of sources. Their reactions to these influences are especially apparent in the Bourbon dynasty's changes to trade and emigration policies as well as shifting social perspectives on race and status during the 1700s.

Economic Heritage: Trade and Emigration

Spain has a long and varied history that includes numerous invasions and colonization periods as well as the convergence of Islam, Judaism, and

Christianity. By 1492, when Cristóbal Colón landed in the Americas, Fernando and Isabel had united the kingdoms of Castile and Aragon and successfully expelled the last of the Moorish rulers in Grenada, bringing the entire Iberian Peninsula under Christian control and concluding an eight-hundred-year war known as the Reconquista. The two hundred years following this historic event are known as the Siglo del Oro, or the Golden Age, of Spain and have been the subject of numerous historical studies (e.g., Elliott 1963).

While many of these studies consider Spain as a whole, the autonomous region known today as Andalucía, located on Spain's southwest coast, is perhaps one of the most important areas for studying Spanish Atlantic history (Figure 2.1). In addition to having an incredibly rich archaeological heritage, the area has deep historical, cultural, and economic connections to the Americas that date to the beginning of Spain's expansion efforts.

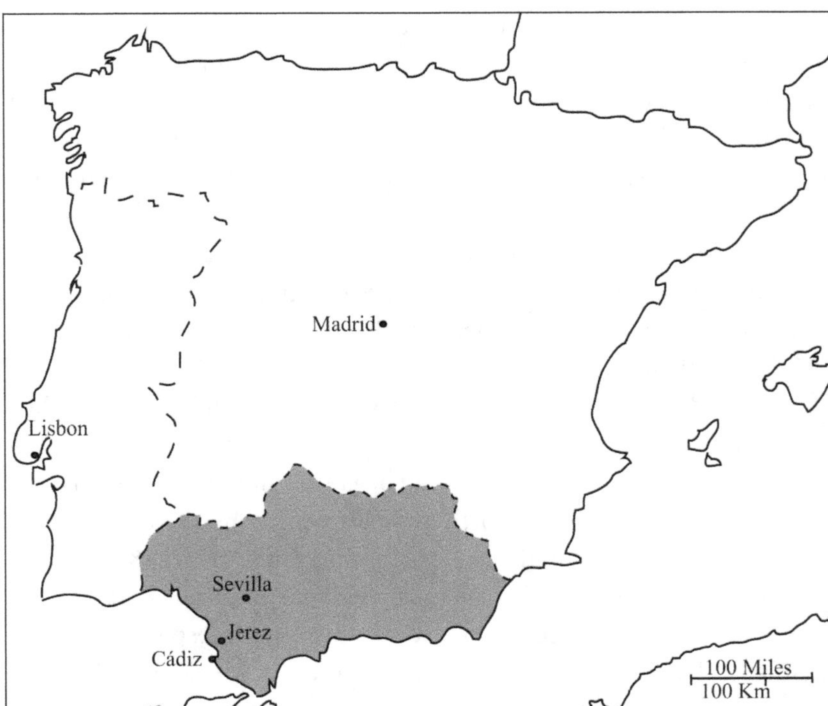

Figure 2.1. Map of Spain with Andalucía shaded. Map drawn by the author.

Trade

In 1503 the Spanish Crown established the House of Trade, or Casa de Contratación, in the Andalucían city of Sevilla, the largest and wealthiest city in Castile (Haring 1918:7–8; Piernas Hurtado 1907:18; Plaza Prieto 1976:410). By 1525 all individuals and goods bound for the Americas had to be registered at the Casa de Contratación prior to crossing the Atlantic (Deagan 1983:20; Elliott 1963:182; Friede 1951:333, 339, 345–347). The location of the Casa de Contratación and the volume of trade that passed through it proved to be a significant support to the local economy (Deagan 1983:20; Haring 1918:8), as will be discussed in greater detail.

For Spanish American customers, however, this mercantilist system proved to be ineffective. Between the mid-sixteenth and the eighteenth centuries, the Crown required that shipments to the Americas be sent via heavily guarded and expensive convoys, such as the *flota*, in an attempt to regulate and protect trans-Atlantic shipments from privateers and pirates (Haring 1918:201–230; Skowronek 1984a). Additionally, because all ships traveled together, the convoy's frequent delays meant that it could sometimes take a year or more for goods to get across the Atlantic (Haring 1918:208–209). For Spanish Americans who did not live on the main trade routes, or Carrera de las Indias, these regulations made it difficult to legally obtain European objects, often forcing them to turn to the *rescate*, or illegal channels, for supplies (Deagan 1983:34–35, 36–37; Ewen 1991:27).

St. Augustine, Florida, for example, was removed from the main *flota* route and therefore received official Spanish supplies from Mexico via the *situado*, an annual government subsidy (Deagan 1983:34–35; Deagan 1987:22; Harman 2004:13). As with the *flota*, the *situado* was frequently delayed, and citizens of the Florida town often complained that they were without the European comforts they deemed necessary. Ostensibly to compensate for this shortage, they began trading with their British-American neighbors to the north as well as engaging in acts of piracy themselves (Harman 2004:13, 36, 51–52; Waterbury 1983:79–81). Despite these supply problems, archaeological evidence indicates that Spanish Americans had a strong preference for Spanish goods, especially ceramics, during the sixteenth and seventeenth centuries, which has led several scholars to conclude that these items were used as a means of expressing social and ethnic status (Deagan 1983; Ewen 1991).

With the advent of the Bourbon dynasty in the eighteenth century, the new monarchs instituted several key changes to Spanish Atlantic trade that had a significant impact in both Spain and the Americas. Between 1717 and 1720, Felipe V (reigned 1700–1746) transferred the Casa de Contratación to Cádiz and restructured its operations, including reducing taxes on exported goods to the Americas (Deagan 1987:22; Gibson 1966:168; Plaza Prieto 1976:411). The trade center's relocation resulted in an economic boom in the region around Cádiz (Elliott 1963:297; López Rosendo 2005a:285). The sudden prosperity and new opportunities for trade in the area are evident in towns like Jerez de la Frontera, which saw a substantial increase in building activities and imported ceramics in the eighteenth century (Aroca Vicenti 2002; López Rosendo 2005a).

While these changes helped improve efficiency and resolved some of the economic problems that plagued Spanish Atlantic trade during the Habsburg reign, Carlos III (reigned 1759–1788) continued to expand the reforms by reducing taxes and allowing other Spanish ports to trade with the Americas. Despite the increase in competition, Cádiz and its neighboring towns, including Jerez, continued to prosper (Plaza Prieto 1976:443).

Although the new trade regulations meant that all American ports were allowed to receive shipments from Spain, smuggling, such as the *rescate*, continued to be rampant in the Caribbean as British, Dutch, and French privateers, pirates, and traders repeatedly violated the trade restrictions, either by commandeering goods meant for the Americas or by bringing in their own cargo to trade with Spanish Americans. It was not until 1793 that foreign trade was openly permitted in Spanish colonies; even then, however, the trade was required to be conducted in Spanish ships (Deagan 1987:23).

It is important to remember that these illegal trading activities do not account for all of the non-Spanish, European goods found in Spanish America. While all ships were required to register at the Casa de Contratación before crossing the Atlantic, their cargo was not necessarily from Sevilla or Cádiz. Shortly after establishing settlements in the Americas, Spain was unable to supply the demand for goods both at home and abroad (Elliott 1963:195–196; Haring 1918:113). During the Habsburg reign, for example, nearly five-sixths of the goods going to the Americas were of non-Spanish, European origin (Deagan 1987:20; Haring 1918:113).

By the early eighteenth century, these foreign goods included fabrics, brass weights and cooking vessels, wax, tin plates, and soap from Flanders, Holland, England, France, Hamburg, Italy, and Portugal (Stevens 1702). Documented Spanish products being sent to the Americas were largely foodstuffs, including wheat, barley, wine from Jerez, salt, and saffron, as well as cloth, wax, and iron (Stevens 1702). While the majority of the items included on these lists are not recovered archaeologically, the global nature of Spanish Atlantic trade is reflected in the American archaeological record by the diverse array of ceramic types, including vessels from England, British America, Mexico, Spain, Portugal, France, the Netherlands, and Asia.

In addition, a wide variety of goods from Spanish America were sent to Europe, including pearls, precious gems, silver, gold, wood, hides, cotton, spices, cochineal, and other raw materials (Stevens 1702). Although historical accounts focus on the more precious commodities that were exported from the Americas, the archaeological record indicates that more mundane objects like ceramics were also shipped to Europe.

Emigration

While Spanish trade and economic policies have been studied in depth for decades, emigration patterns, especially during the eighteenth century, are relatively less well known. Archives throughout Spain and the Americas, especially the Archivo General de Indias have yielded a wealth of information regarding Spanish Atlantic migration; studies of these documents, however, reveal that the eighteenth-century documentary record is often incomplete and accounts only for legal, recorded emigration and does not factor in illegal emigrants (Mörner 1976).

Despite these issues, research to date suggests that, in many ways, the emigration patterns from Spain to the Americas mimicked the trade activities described above. Initially, only individuals from Castile, which included Extremadura, Pais Vasco, and Andalucía, were allowed to travel to the Americas. Because all emigrants were required to register at the Casa de Contratación in Sevilla, the majority of these individuals were from Andalucía (Boyd-Bowman 1976:582, 589–591; Deagan 1983:20–21). While early emigrants were primarily unskilled young men, an increasing number of professional men, artisans, women, and families moved to the Americas by the mid-sixteenth century to take advantage of the region's affluence and escape Spain's economic troubles (Boyd-Bowman 1976:582–584; Mörner 1995:253).

By the eighteenth century, emigration patterns had shifted in several significant ways. Fewer professionals or soldiers were traveling to the Americas, as nearly 70 percent of Spanish emigrants were *criados*, or servants (Mörner 1995:253). Also, the number of individuals from Andalucía arriving in the Americas decreased dramatically as more individuals emigrated from northern regions of Spain, such as Galicia and Catalonia, as well as the Canary Islands (Mörner 1976:755; 1995:253, 257, 260).

Despite the increased diversity of emigrants, the Andalucía–America connection most likely remained fairly stable during the eighteenth century. Letters from the period indicate that many Spanish Americans were in frequent contact with their family members in Spain and often encouraged them to join them in the Americas (Lockhart 1976; Mörner 1995:264). In some instances, individuals returned from the New World to Spain, although statistics on the number and demographics of this group remain unclear (Lockhart 1976:784; Mörner 1976:749–750). Similarly, the Cádiz region continued to be a prominent player in trans-Atlantic trade and thereby would have been in a prime position to disseminate fashions, news, and sociocultural developments.

Social Heritage: Status and Identity

Although the trans-Atlantic trade and emigration helped forge and preserve strong economic and social connections between Andalucía and the Spanish Americas, they did not exclude the influences of other regions or cultures. Items from Asia and other parts of Europe and interactions with both Africans and indigenous populations were prevalent in both Spain and Spanish America. While these influences sparked a social struggle in Spain about what it meant to be Spanish at the beginning of the Bourbon dynasty (Fattacciu 2011:154), the issue of cultural and ethnic purity was by no means limited to the eighteenth century.

Limpieza de Sangre

Initially, this debate was closely connected to ethnic and religious heritage. As hinted at by Don Quixote in the quote at the beginning of this chapter, status was closely tied to one's lineage and whether or not one came from a noble dynasty or a family that had recently risen to greatness.

When the Moors invaded the Iberian Peninsula in 711, the idea of "race" became a way of distinguishing among Christians, Moors, and Jews

(Carrera 2003:10). After the Reconquista in 1492, this distinction became a means of identifying non-Christians, many of whom were discriminated against and persecuted (Carrera 2003:10; Pike 1972). Ultimately, the Spanish Crown instituted a series of decrees known as the *limpieza de sangre*, or "pure blood," in which individuals were categorized by their heritage and, if they were not Christian, were forced to either nominally convert to Catholicism or leave the kingdom.

During the eighteenth century, as Enlightenment scholars sought ways to organize and classify the natural world, this preoccupation with religious and ancestral heritage became an increasingly important aspect in determining one's place in society (see Carrera 2003:1–21). While sixteenth-century Spaniards focused on one's religious background, eighteenth-century Spaniards looked to one's *raza*, or biological lineage (Carrera 2003:10; Kuznesof 1995:160). An individual's lineage, in turn, enabled others to identify his or her *calidad*, or status, and therefore classify him or her in the broader social order (Carrera 2003:xvii, 10).

Casta System

The preoccupation with status and ancestral heritage traveled with the colonists across the Atlantic, where relations among Europeans, indigenous people, and Africans produced a complex social melting pot. In Spanish America, one's social standing was largely dependent upon the degree of Spanish blood one could claim. Initially, Spanish American society was divided into two main groups: the República de los Indios and the República de los Españoles (Carrera 1998:38). These republics, in turn, divided people into three additional categories based on their heritage and whether they had Spanish, indigenous, or African ancestors (Carrera 1998:38). Tellingly, Africans were not included in either of the two republics but instead were seen as answers to a labor problem rather than as major contributors to racial or social demographics (Carrera 1998:38).

Predictably, the situation in the Americas quickly became more complex as intermarriage and interbreeding between the groups blurred these neatly defined categories. The combination of the three primary groups initially formed three additional categories, or *castas*: Spanish Indian (*mestizo*), Spanish African (*mulato*), and African Indian (*zambo*) (Seed 1982:572). By 1540, the crown had established between 14 and 20 distinct *castas* in order to accommodate and identify children of parents that fit within the original groups described above (Carrera 1998:38).

Yet another set of divisions was introduced in the seventeenth and eighteenth centuries to identify one based on his or her birthplace or skin color. Spaniards born in Spain, for instance, were called *peninsulares*, while those of purely (or seven-eighths) Spanish descent who were born in the Americas became known as *criollos* (Shephard 1983:65). Additionally, a light-skinned *mestizo* individual was called a *castizo* (Seed 1982:573). A light-skinned *mulato* individual, in contrast, was known as a *morisco* (Seed 1982:573), not to be confused with the *moriscos* of Spain who were Moors converted to Christianity. Interestingly, there does not appear to be a specific category for non-Spanish European individuals within this *casta* system, despite their being present in Spanish America.

In eighteenth-century Spanish America, this lineage-based social structure was perpetuated through *casta* paintings, or sets of images that depict idealized, fictitious families representing the various *casta* groups. While nearly all of the images include textual identification of the *castas* being illustrated, most also place each family group in a specific setting with the appropriate clothing and material culture for the family's category (Loren 2007). The *castas* were organized hierarchically with Spaniards and individuals of Spanish descent on one end of the series, and those of less pure European heritage at the other end (Carrera 1998:43; Kellogg 2000:74).

For a largely illiterate audience, these images can be seen as a way of disseminating ideas regarding appropriate behavior, clothing, and possessions for the different social groups as well as educating people in Spain about the social diversity in the Americas. The importance of external attributes in identifying one's *calidad* is evident in Mexican court documents that use nonbiological characteristics, such as clothing, jewelry, behavior, speech, housing, and friends, to determine an individual's *casta* (Carrera 2003:5).

Spanish Identity

This preoccupation with classification and identity is predominant at several points throughout Spanish history and can be seen as a reaction to significant social and political changes within Spain and her empire. Felipe V's ascension to the throne as Spain's first Bourbon monarch in 1700 marked the beginning of one such crisis. Prior to this point, Spaniards had been relatively isolated from developments in other parts of Europe, and certain books and ideas had been banned under the Inquisición

(Herr 1958). This seclusion largely ended with the arrival of a new king, who had lived in France his entire life up to his coronation, and his subsequent appointment of a French and Italian court. These influential foreigners popularized French fashions, cuisine, and technology, while also introducing Spaniards to Enlightenment philosophy and non-Spanish European ideas, such as the works of Voltaire and Linnaeus (Herr 1958).

By the 1760s Spanish society was divided over how to react to these new influences. Some embraced the new fashions and ideas, while others felt that they diluted and degraded Spanish culture. In many ways both sides expressed their views through their material culture, especially by the adoption or rejection of French clothing. Those who embraced these new trends sought to imitate the French-inspired court fashions, both in clothing and behavior; the elites who rejected and resented the changes began to dress like the lower classes, because they felt these fashions embodied traditional Spanish culture (Martínez Bermejo 2008; Noyes 1998). The issue quickly escalated beyond a social and sartorial debate, as can be seen in revisions to sumptuary laws, which regulated personal spending in an attempt to ensure that only those with upper-class social standing owned and used luxury goods.

In 1766 Carlos III modified these laws to restrict upper-class individuals from dressing and acting "below" their ascribed status and therefore "tarnishing their rank" (Noyes 1998:209–210). This shift is telling in that for the first time these laws regulate upper-class behavior rather than restrict the activities and consumption practices of the lower classes. In keeping with tradition, these regulations appear to have been largely ignored by both officials and the general public; elite individuals continued to copy peasant fashions and, in Madrid, even rioted against a decree issued by the Marqués de Esquilache that banned traditional broad-brimmed hats and long capes. Although some scholars have attributed these acts of rebellion as upper-class opposition to the perceived invasion of Enlightenment ideals and French fashions (Martínez Bermejo 2008; Noyes 1998), they are also indicative of the intense struggle during this period to define what it meant to be "Spanish" while also negotiating a changing, more global world (Fattacciu 2011:154).

This identity crisis continued into the nineteenth century when, in 1808, the Mexican Revolution signaled the beginning of the end of the Spanish Atlantic empire, and again in 1898, when Spain lost Puerto Rico, Cuba, and the Philippines and was defeated by the United States in the

Spanish-American War (Vásquez 1995:85). By the 1890s there was a movement to revitalize Spanish society, although again there were two schools of thought as to how to accomplish this task (Vásquez 1995:86). One side of the debate, known as the "generation of 1898," argued that Spain's problems and decreasing global influence stemmed from a continued sense of isolation from Europe and the rejection of European ideas (Vásquez 1995:86). Those opposed to this view, like essayist Miguel de Unamuno, championed a movement known as *casticismo*, which held that Spaniards needed to reexamine their past to discover what was quintessentially Spanish, specifically Castilian (Krause 1993:19; Unamuno 1902; Vásquez 1995:86).

Both the riot in 1766 and the 1898 debates show a recurring preoccupation with the definition of being "Spanish" and how or even whether to incorporate new cultural influences. It is interesting to note that Spaniards did not react as strongly to changes that came from the Spanish Americas, including new foods and ceramics, a rising middle class because of an influx of new wealth, and new social categories like the *casta* system.

For archaeologists, the use of material culture to indicate identity is a well-known practice (Deagan 1983; Ewen 1991; Loren 2000, 2007, 2008; Rothschild 2006; Voss 2005, 2012; Wilkie 2000). In early modern Spain and Spanish America, public appearance was typically used as a basis for identifying a person's status. Clothing, however, is easily changed and can be manipulated by individuals to move up or down the social scale or even to assume different identities in different situations (Donahue 2004; Loren 2000, 2008; Noyes 1998; Voss 2008a). As Spanish literature scholar Darcy Donahue (2004:106) notes, "in a culture of appearances such as that of the early modern Spanish city, attire could provide socioeconomic readability, although an increasingly unreliable one." In *Don Quixote*, for example, a wealthy-woman-turned-shepherdess dresses the part of a shepherdess and lives in rural solitude; she contradicts her physical appearance and indicates her true status and breeding, however, when she gives an eloquent speech justifying her reasons for abandoning her inherited wealth and social status (Cervantes 1605 [1964]).

Because clothing and publicly-visible accessories were often unreliable social markers, other, more dependable markers would have been necessary to fully demonstrate identity and status. While appearing wealthy and socially elite on the street might have been possible by simply dressing the part, the most effective demonstrations of status and social aspirations would have involved owning specific items, such as luxury tablewares, and using them correctly, such as in fine dining with the appropriate foods and dishes (Beaudry 2008:191). In other words, to be convincing, one would need to demonstrate his or her status not only through clothing and accessories but also through proper behavior and household items.

Dining and entertaining, therefore, are prime examples for studying these alternative expressions of social identity. When conducted in a household setting, these activities would have been more intimate than in public spaces like the street or market, but still socially visible to nonfamily members. By inviting nonrelatives into the house, the family would be given the opportunity to display wealth and status in ways not normally possible when passing the same people in public. Consequently, luxury ceramics and tableware, an extralegal category that was not regulated by the sumptuary laws and was common on both sides of the Atlantic, provided individuals with alternate ways to express their social goals and cultural identity, specifically through semipublic domestic activities.

Conclusion

Heritage and status were of extreme importance to early modern Spaniards. On both sides of the Atlantic, individuals used material culture to convey and manipulate their *calidad* or *casta* status. In eighteenth-century Spain, people often used clothing to express either their knowledge of current trends in the Bourbon court or their commitment to "traditional" Spanish culture. Individuals in the Americas, in contrast, relied on trade and emigration to identify and obtain fashionable goods that expressed their connections to the broader Spanish Empire and their position as Spanish citizens.

While clothing and objects were key to demonstrating one's status, they could be easily changed, especially if one used clothing or other material goods to manipulate his or her public persona for economic, personal, or social reasons. Consequently, it is important to consider both the objects and the behavior that accompanied the items, especially with regard to dining and entertaining. Ceramic vessel forms offer an ideal means of considering social behavior, because they provide evidence of specific activities and dining practices while also reflecting changes in fashion and foodways.

3

Ceramic Organization in the Spanish Atlantic (COSA)

In the seventeenth century, Spaniards like Don Quixote were preoccupied with how to classify people according to their social, ancestral, and ethnic backgrounds. They also had a strong sense of how to identify and catalog the objects they encountered in daily life, including ceramics.

When one does not abide by a culturally established classification system and sees objects differently from those around him or her, an individual is often unable to effectively communicate with others and can be perceived as nonsensical, as was Don Quixote. At one point during his adventures, for example, the Knight of the Rueful Countenance encounters an item that he determines is the enchanted helmet of Mambrino, a fabled knight-errant whom he admires. While Sancho sees the item for what it is, a brass barber's basin, Cervantes's hero explains that the object had been transformed to look like a basin but that it would still protect its wearer from any blow (Cervantes 1605 [1964]). This basin then becomes his chosen headgear and a prized item that he protects and carries with him in subsequent adventures.

To understand a culture, it can be important to take Sancho's example and see objects for what they are to their users and makers, rather than following the model set by Don Quixote and risk creating a possibly inaccurate interpretation of items. For decades archaeologists have attempted to find appropriate ways of classifying the objects they recover, in order to accurately interpret their findings. Regardless of the importance of organizing and identifying objects, classification systems and typologies remain a confusing and, at times, controversial issue.

This situation is especially problematic in the Spanish Atlantic, where scholars have different approaches to archaeology and have developed different artifact classification systems and analysis. In Europe archaeology is often closely allied to the field of history, as the former was initially developed in an attempt to extend knowledge of the past beyond the written documents used by historians (Trigger 2006:248). While Spanish archaeology has adopted theoretical and methodological approaches from Central Europe, England, and America (Vázquez Varela and Risch 1991), it retains its strong links to history and is often considered a scientific approach to studying history. Indeed, many Spanish universities currently offer archaeology courses in departments that combine history, prehistory, geography, and archaeology.

In the Americas, in contrast, archaeology is considered a subfield of anthropology, in part because of the influence of Boasian cultural anthropology during the field's early development (Longacre 2010; Trigger 2006:278). As a result, American archaeological research often focuses on understanding the past from an anthropological perspective and exploring people and cultures (Binford 1962; Longacre 2010). While archaeology's role in anthropology is sometimes questioned, it remains under the umbrella of anthropology departments in the majority of universities in the United States, Mexico, and Canada (Longacre 2010).

Unfortunately, these differences have meant that the data from the two sides of the Atlantic are often incompatible and that scholars have been unable to compare the research being conducted on Spanish and Spanish American sites. Such comparisons, however, have a potentially big interpretive reward, because they allow us to investigate changes within the Spanish Atlantic empire and to consider cultural developments on both sides of the Atlantic in relation to each other.

These developments are particularly evident through ceramics, which were a ubiquitous element of material culture throughout the Spanish Atlantic. Current historical and documentary research, for example, has identified over 150 different Spanish terms for ceramics (Amores Carredano and Chisvert Jiménez 1993; Lister and Lister 1976; Ramos Palencia 2010; Ruiz Gil 1999). People were surrounded by pottery, ranging from tiles on fountains and walls to cups and plates on tables to *bacines* (chamber pots) in private areas. Pottery also constitutes a significant amount of the archaeological record. At La Calle Corredera, for example, nearly 85 percent of the nonorganic objects inventoried are ceramic.

Unfortunately, it is difficult to compare ceramic assemblages from Spain and the Americas because of different research approaches. In Spain archaeologists focus on vessel form more than on ceramic type, and when they need to reference a specific kind of ceramic, they often rely on an American-based typology that does not always fit their data. One possible reason for this methodology is that, until recently, archaeologists have not focused on early modern sites and therefore have not yet developed a standardized typology for the plethora of ceramic types produced in early modern Spain. Scholars working in the Americas, in contrast, use a detailed typology based on ceramic paste, glaze, and decoration, and they do not always publish information regarding vessel forms. Both of these systems have their strengths. The Spanish, vessel-based approach sheds light on how the objects were used, while the American, archaeological typology provides data on ceramic industry, chronology, and trade.

Ceramic Organization in the Spanish Atlantic (COSA)

To compare the Spanish and Floridian sites, this study employs a vessel-based folk classification system known as Ceramic Organization in the Spanish Atlantic (COSA) (Ness 2015a), which includes data from both sides of the Atlantic. In both modern and eighteenth-century Spanish, the term *cosa* literally translates to the word "thing," making this acronym especially appropriate for a classification system that focuses on the object or vessel rather than on the glaze, decoration, or paste type. This emphasis on vessel forms makes it possible to consider how the objects were used and perceived by early modern consumers and, by extension, what activities were being performed at a given site.

COSA, which is inspired by the Spanish classification system proposed by Fernando de Amores Carredano and Nieves Chisvert Jiménez (1993) and the Potomac Typological System (POTS) (Beaudry et al. 1983), is organized on three levels: the quality, the purpose or intended function, and the form itself. At the broadest level, the forms are grouped according to categories defined by early modern Spanish and Mexican guilds that divided potters based on the quality of their products, namely, *común,* or low-grade utilitarian or service wares, and *loza fina,* or high-quality tablewares. Within these two quality-based groups, the forms are organized according to their intended purpose. The smallest unit in COSA is the form itself, with the most generic or commonly used term to represent specific forms.

In order to ensure an emic (user-ascribed) perspective, the terms have been selected from a variety of documentary sources, including analyses of probate inventories, ceramic glossaries, and early modern dictionaries and cookbooks. This combination of sources, as well as archaeological data and contemporary paintings, has provided a wealth of detail regarding the shapes, how the objects were used, and whether or not the same shapes were made from other materials.

It is perhaps worth noting that, although the documentary sources date to the eighteenth century, the terms themselves seem to have remained relatively consistent since the sixteenth century. While the majority of earlier texts do not typically include detailed descriptions, the first standardized Spanish dictionary, *Diccionario de autoridades* (1726–1739), does include extensive descriptions as well as older terms that were used by fifteenth- and sixteenth-century authors (Pountain 2000:167–169). As a result, COSA, although designed to facilitate eighteenth-century research, should also be useful for scholars working on earlier sites. Indeed, while the systematic framework and definitions below are new, archaeologists working with Spanish ceramics have long used many of these terms and will hopefully be comfortable applying this system to their own work.

Likewise, while this book applies COSA to Spanish Atlantic sites, the fairly standardized nature of Spanish terms for ceramics suggests that this system could also be useful for those studying other portions of the Spanish Empire, including the Philippines. Regional differences may exist, such as the Andalucían term *pocillo* (chocolate cup), as opposed to the more broadly used term *jícara* for the same vessel; however, eighteenth-century documents appear to use a consistent vocabulary, and contemporary dictionaries often reference more obscure terms, thereby helping the modern researcher account for regional differences. For the sake of consistency and to ensure that COSA is useful to a broad audience, this system relies on common Castilian terms, although some regional variations that might appear in documents are also noted.

In addition to dictionaries, cookbooks proved especially informative regarding connotations, frequency, and use of different vessels. As other scholars have noted, cookbooks and other texts often indicate unexpected ways in which vessels might have been used and highlight the multipurpose nature of objects (Scott 1997). Glasses or teacups, for example, were recommended for use as cups, cookie cutters, measuring cups, and so on (Scott 1997). Even Don Quixote's "Mambrino's helmet" seems to have had multiple uses, as the barber who originally owned it used it both as intended and as waterproof headgear. While this latter use of a barber's basin is doubtless uncommon, it does illustrate that people sometimes used their objects in unexpected ways that might not be obvious from an archaeological perspective.

In the most common eighteenth-century Spanish cookbook, *Nuevo arte de cocina*, terms also tended to be multipurpose. Recipes were often called *escudilla* or *plato*, presumably depending on the kind of vessel in which the food would be served. Stews and soups, for example, were frequently referred to as *escudillas,* and this term also served as a liquid measurement (Altamiras 1758). Similarly, non-broth-based dishes, such as *ranas en pastelillos*, or "frogs in cakes," were often described as *platos* (Altamiras 1758).

Although the multipurpose nature of terms can make it tricky to use historic texts for material culture studies, it also indicates the importance of vessel forms for early modern Spaniards and the value of considering this approach when classifying ceramics. An investigation of vessel forms can shed light on Spanish foodways from a practical standpoint, as frequent references to *ollas*, or pots, for example, suggest a stew-based diet. From an emic perspective, this approach indicates important connections between food and vessels. The individuals using the ceramics in question saw these objects in terms of function rather than decoration, and they frequently relied on the same terms to identify objects, functions, and contents.

It is important to remember that COSA was designed specifically for ceramics found on early modern (sixteenth–eighteenth century) Spanish Atlantic domestic sites. It therefore includes vessel forms commonly recovered from Spanish and Spanish American households that can be identified archaeologically. Because other sources (Amores Carredano and Chisvert Jiménez 1993; Goggin 1960; James 1988; Lister and Lister 1987) address vessels intended for industrial, agricultural, or transportation purposes, including olive jars, these shapes are not included in this classification system.

Loza Común

FOOD PREPARATION

Cazuela A casserole or wide earthenware vessel used primarily for stewing or cooking meats, fish, eggs, or other foods over a fire without other ingredients or a sauce (Lister and Lister 1976:33; Real Academia Española 1726–1739, 1780). Unlike the term *olla*, *cazuela* seems to have been used primarily for the vessel form and not for the foods cooked in it. One sample from La Calle Corredera in Jerez measured approximately 30 cm in diameter, although this form might have come in multiple sizes.

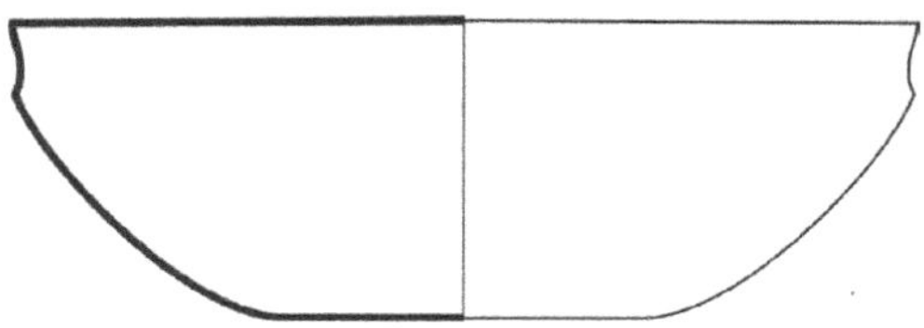

Mortero A mortar used to crush spices, seeds, or drugs, as well as a large or porous stone used to filter and sanitize water (Real Academia Española 1726–1739). Eighteenth-century cookbook references to *morteros* typically mention crushing garlic or sugar (Altamiras 1758), although the vessels also might have been used for pharmaceutical purposes (Lister and Lister 1976:63). While eighteenth-century dictionaries state that *morteros* were typically made of stone or wood, at least one ceramic mortar has been recovered at La Calle Corredera.

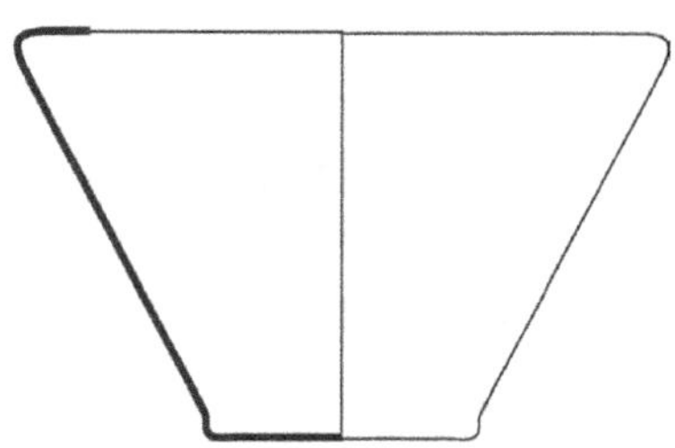

Olla	A cooking pot described as a round earthenware vessel. Eighteenth-century dictionaries are very specific about this shape, stating that the vessels have a narrow base, wide belly, slightly tapered mouth, and at least one handle (Lister and Lister 1976:66; Real Academia Española 1726–1739). While dictionaries indicate that this term also sometimes refers to the stew made in these vessels, cookbook authors often use this word to refer to the vessel, not its contents (Altamiras 1758; Real Academia Española 1726–1739). *Nueva arte de cocina* also uses the term *puchero*, referring to a lead-glazed vessel that was smaller than an *olla* but essentially used for the same purposes (Altamiras 1758; Real Academia Española 1726–1739). 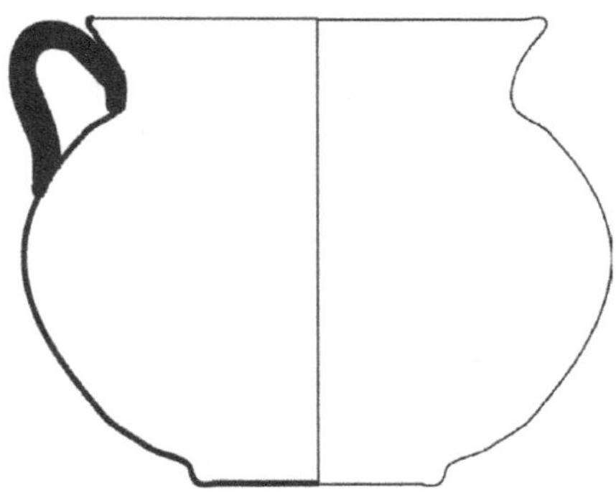
Orza	A tall, glazed earthenware jar without handles that was used for storing preserves or drugs (Lister and Lister 1976:66–67; Real Academia Española 1726–1739). A partially complete example was recovered from La Calle Corredera that measured approximately 16 cm tall and was glazed on the interior, with minimal decoration on the exterior. 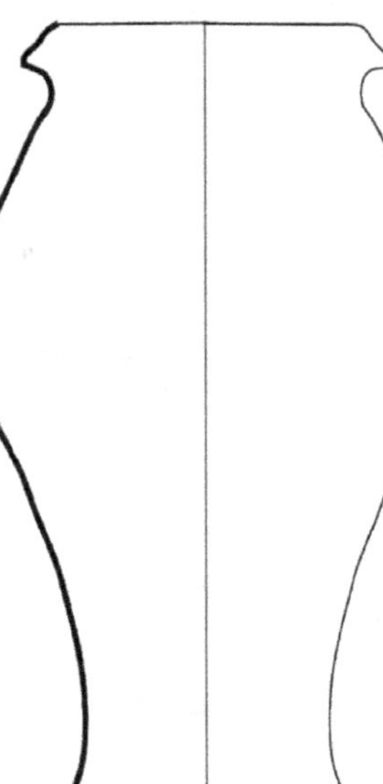

(*continued*)

Lebrillo A round, glazed basin with flared sides and a wide mouth (Real Academia Española 1726–1739, 1780). Although dictionaries identify this form as a glazed vessel, decorations on archaeological examples vary widely, including undecorated, unglazed, incised, or elaborately handpainted and glazed designs. Rim decorations also vary, as some *lebrillos* have a flat rim that has a rectangular profile, while others have a slightly more decorative rim with molded or impressed patterns. All seem to have flat bases and typically undecorated exteriors.

As a vessel form, *lebrillos* appear to have been all-purpose vessels, with uses ranging from preparing and serving food to bathing and laundry. The term is sometimes spelled *librell* and is often used interchangeably with *cuenco, bacín*, and *brasero* (Lister and Lister 1976:56), although these terms appear to have had distinctly different definitions.

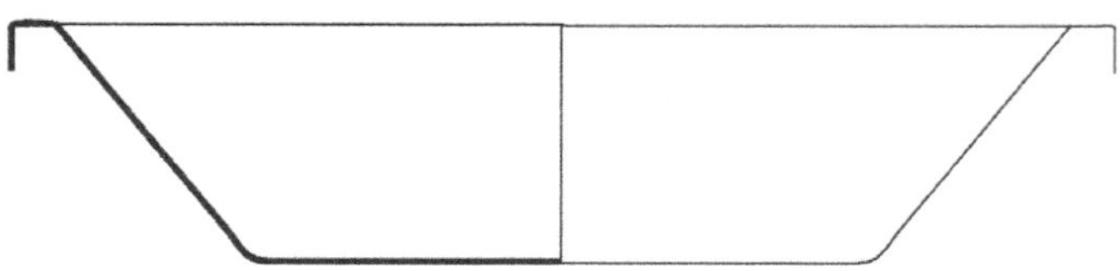

Anafe A portable stove made from iron, earthenware, or soft stone and used most commonly in Andalucía; different forms were used depending on the material (Real Academia Española 1770, 1780). Small, portable stoves also appear to have been called *braseros* or braziers (Lister and Lister 1976:26–27).

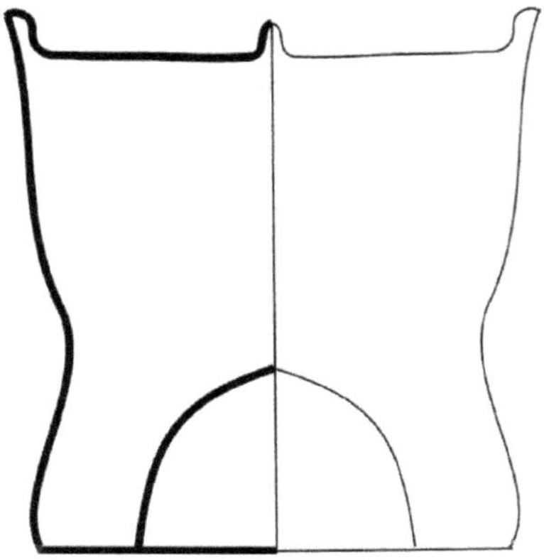

STORAGE AND TRANSPORT

Embudo A funnel used to channel water, wine, and other liquors (Lister and Lister 1976:40; Real Academia Española 1726–1739). While most *embudos* were probably used for liquids, at least one eighteenth-century recipe references using a small *embudo* to make deviled eggs (Altamiras 1758:117–118), possibly to pipe the filling into the hardboiled egg white.

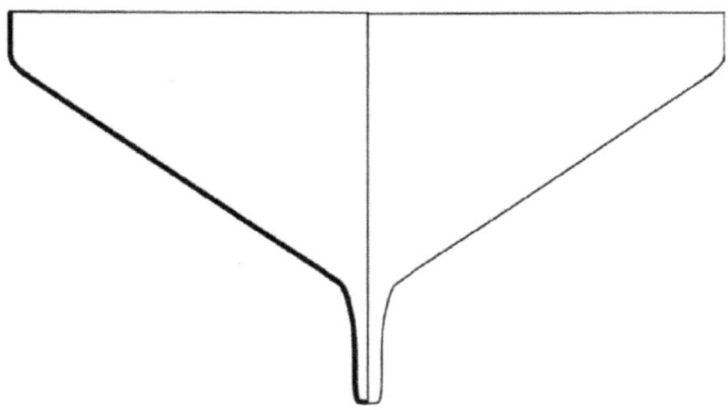

Tina A large storage jar with a wide base and narrow mouth, sometimes described as having a cauldron-like shape (Real Academia Española 1726–1739, 1780). This form is also known as a *tinaja*. Although it might have varied in size and shape, at least one study states that the form has a liquid measurement of approximately $12\frac{2}{3}$ gallons, or 48 liters (Lister and Lister 1976:86).

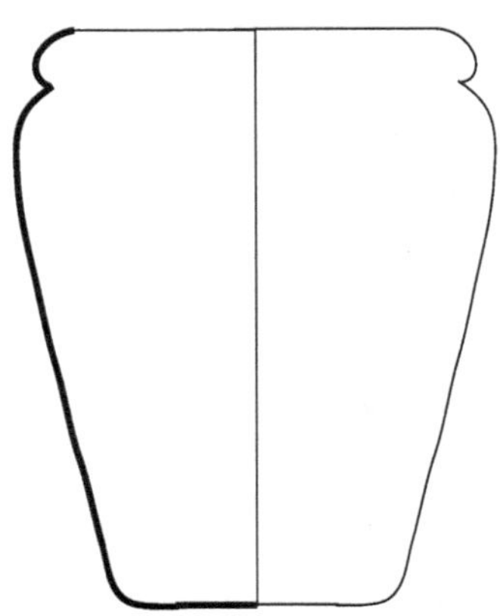

(*continued*)

Arcaduz	An earthenware bucket used to draw water from wells; this term also refers to a pipe for conducting water (Lister and Lister 1976; Real Academia Española 1726–1739, 1780).	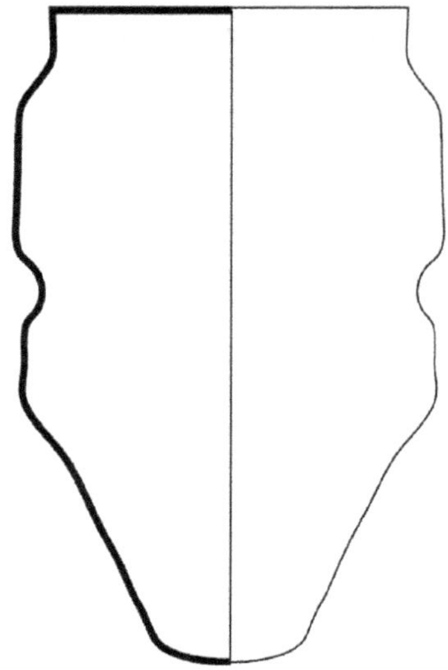
Cántaro	A large earthenware vessel with a wide belly, narrow mouth and base, and one handle (Real Academia Española 1780). These vessels had a capacity of approximately three gallons and were sometimes made of metal (Lister and Lister 1976:31; Real Academia Española 1780). While this term is referenced in *Nuevo arte de cocina*, it was typically used as a measurement for amounts of water that are more than a bowl, or *escudilla* (Altamiras 1758). Some inventories use the term *cantarilla* (Ramos Palencia 2011:47), but this term does not appear in dictionaries until 1803 (Real Academia Española 1803).	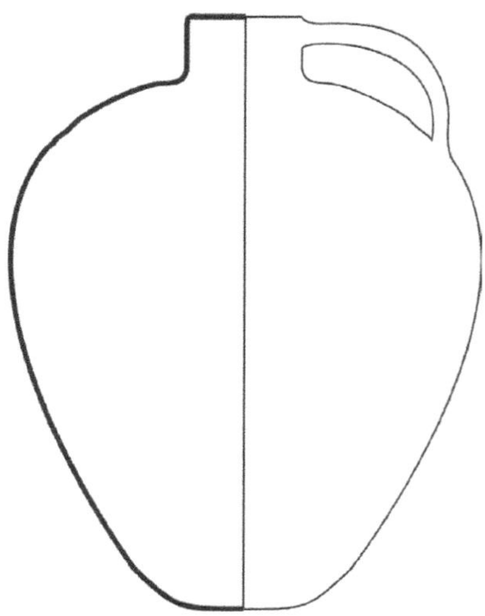

Loza Fina

HYGIENE	
Bacín	A large, cylindrical chamber pot with straight sides and a wide horizontal brim (Lister and Lister 1976:22). Eighteenth-century dictionaries specify that these were round, lead glazed, and specifically for "large" human excrement (Real Academia Española 1726–1739). According to some scholars, the term *bacín* is sometimes confused with *lebrillo* in documents (Lister and Lister 1976:56); *bacines*, however, served a very different purpose from *lebrillos*. 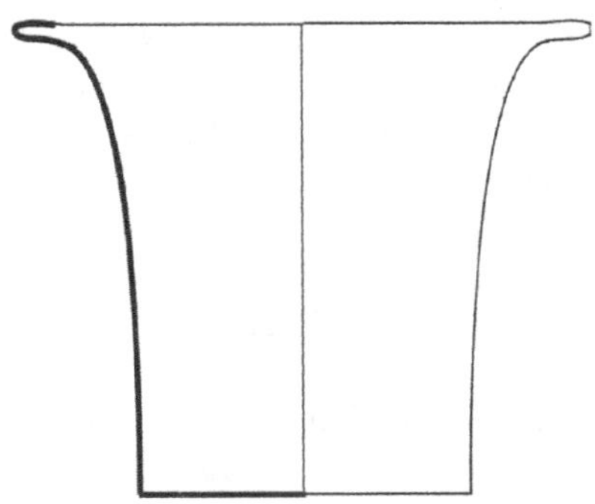
LEISURE	
Pipa de tabaco	A tobacco pipe (Real Academia Española 1780). Eighteenth-century Spanish dictionaries indicate that tobacco pipes could be made from a variety of different materials but do not specify what materials were common. 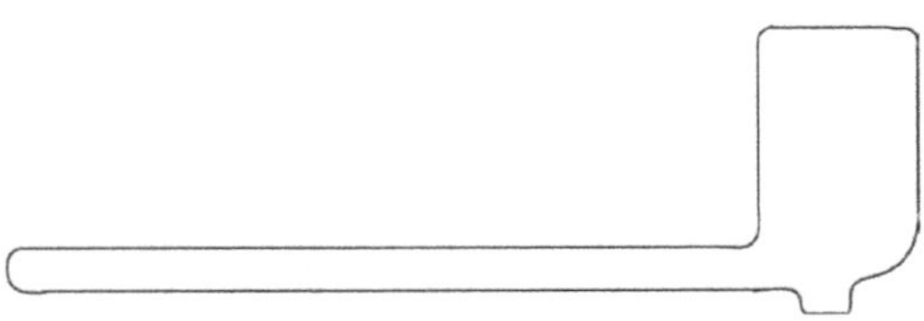
Juguete	A cheap, colorful toy, typically for children (Lister and Lister 1976:54; Real Academia Española 1726–1739). Because this form could have multiple shapes, ranging from miniature cooking vessels like the ones listed above to figurines, it is not possible to accurately illustrate this form.

(*continued*)

Escudilla vieja

A bowl used for consuming soups and broths (Real Academia Española 1780). *Nuevo arte del cocina* suggests that these vessels also might have been used in cooking, especially when determining a serving size or liquid measurement (Altamiras 1758).

Although some archaeologists have suggested that *escudillas* were used for beverage consumption in the sixteenth century (Lister and Lister 1976:41; Skowronek 1991; South et al. 1988), this function does not appear to have been referenced in eighteenth-century documents. This discrepancy suggests that either the use of these vessels changed over time or that one was not supposed to drink from this form. If *escudillas* did serve as drinking vessels, it is possible that this function, like use in cooking, was not considered the primary or "official" function of the vessel and therefore was not documented.

While eighteenth-century dictionaries use the term *escudilla,* archaeological examples of this form suggest that by the 1700s the shape of *escudilla*s had largely changed to a more rounded form, thereby eliminating the straight edges and distinct angle where the body begins to narrow near the base. A partial *escudilla vieja* from La Calle Corredera had a diameter of 11 cm and was approximately 7 cm tall, with a 5-cm-wide foot ring.

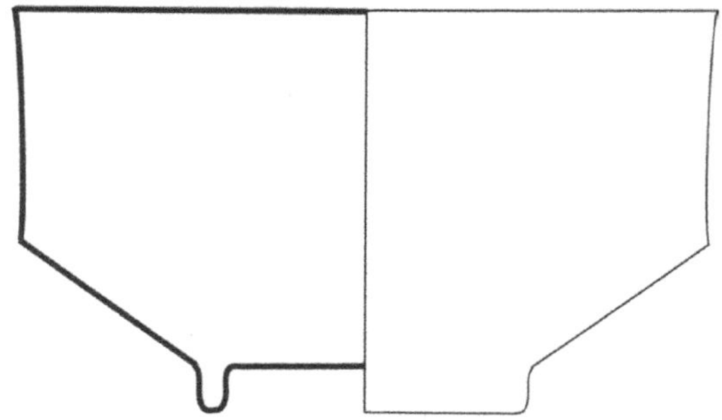

Escudilla nueva	A bowl used for soups and broth that may have been used for cooking, serving, and consuming food or beverages, like an *escudilla vieja* (Altamiras 1758; Real Academia Española 1780; Skowronek 1991; South et al. 1988). Archaeological evidence indicates that this form begins to appear in the eighteenth century and has a rounded body and a foot ring. Samples from La Calle Corredera measured approximately 12 cm wide and six cm tall, making them similar in size to the earlier *escudillas viejas*. This newer form is different from an *escudilla vieja*, however, in that it has a rounder body and lacks an angle near the base. 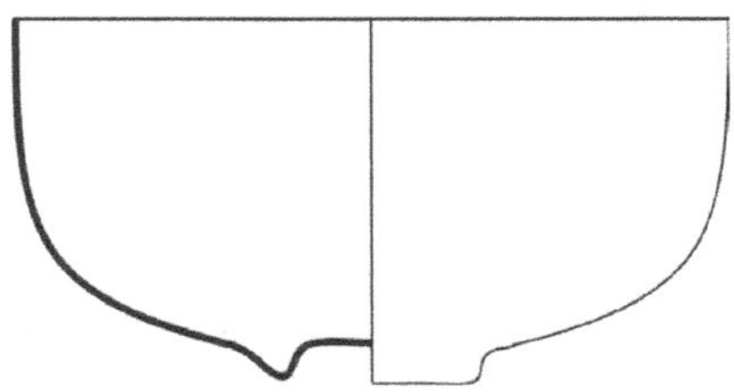
Huevero	An egg-cup on a small pedestal. In eighteenth-century dictionaries, this term is fairly generic in that it refers to anything or any person that deals with or sells eggs (Real Academia Española 1780). 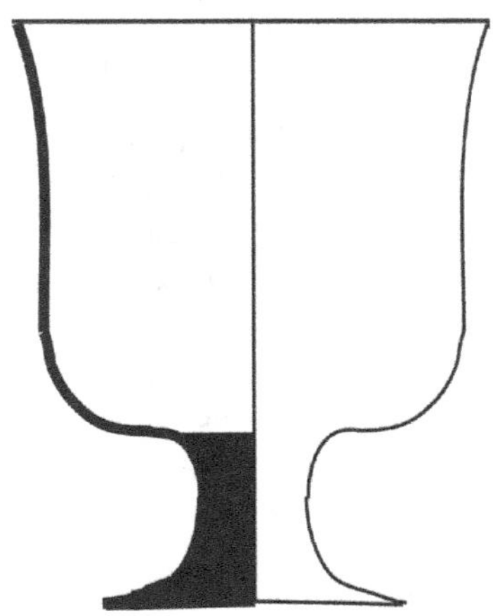
Macerina	A type of plate or saucer with a deep depression or holder in the center to fit a *jícara* to avoid spilling the drink (Real Academia Española 1780). This term is also occasionally spelled "mancerina" (Lister and Lister 1976:58). 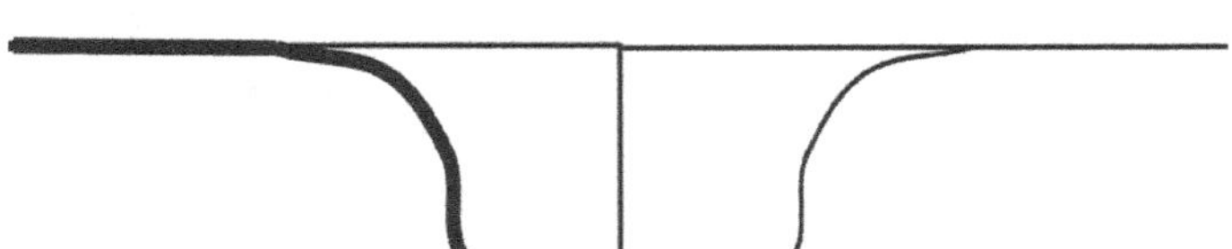

(*continued*)

Plato viejo

A deep plate that is almost bowl-like in shape but with a wide, flaring mouth and occasionally a rim and marly, or flat area next to the rim.

Archaeological examples of this form date to before the 1700s and, in Spain, often measure approximately 22 cm at the widest point. In the Americas, they often have a marly and are therefore sometimes called "brimmed platos" by archaeologists. Because plate forms changed shape in the eighteenth century but kept the same name, the term *plato viejo* differentiates this earlier form from its eighteenth-century successor.

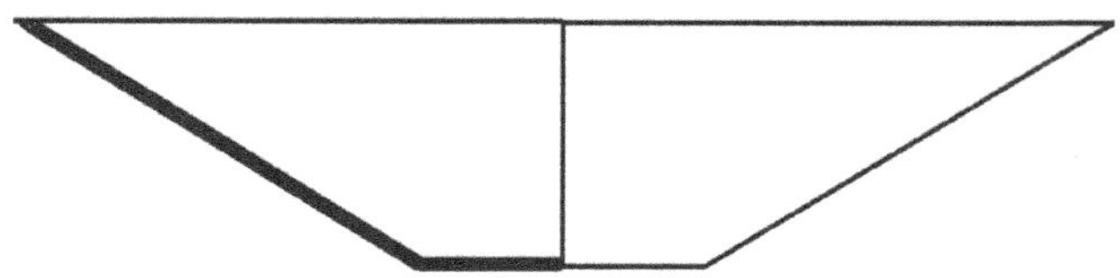

Plato nuevo

A relatively shallow plate with a flat rim, similar to modern plate forms today. Eighteenth-century dictionaries indicate that this form came in multiple sizes and materials, including silver, gold, pewter, ceramic, and wood; these sources specify that it was used for serving and consuming meat, among "other uses" (Real Academia Española 1726–1739).

The form begins to appear in eighteenth-century archaeological contexts and is typically made of a variety of pastes and glazes. While *platos nuevos* range in size, several partial samples were found at La Calle Corredera that measured between 20 and 22 cm in diameter, making them similar in diameter to their predecessors. As mentioned above, the term *plato nuevo* is used to differentiate this form from earlier, deeper plate shapes.

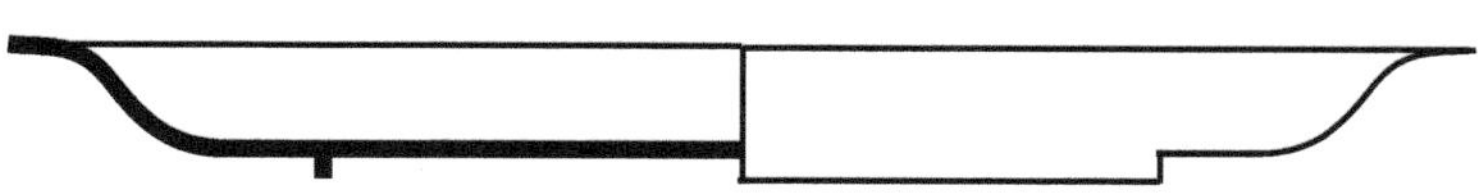

Tapadera	A lid to cover a wide-mouthed vessel, made from a variety of materials (Lister and Lister 1976:83; Real Academia Española 1726–1739, 1780). Archaeological examples from La Calle Corredera measure approximately 11 cm in diameter, are made of a refined buff paste, and appear to match the finer *jarro* vessels from the site. 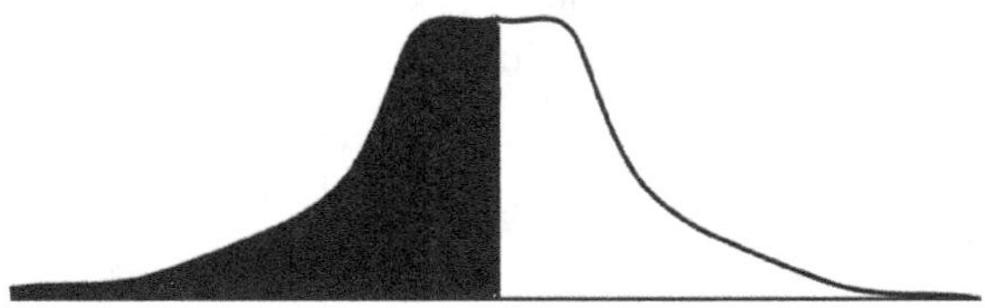
Taza	A cup for drinking and "other uses" (Real Academia Española 1780). Cookbooks from the time period suggest that these additional functions included serving as cooking tools, such as a dumpling mold (Altamiras 1758:110–111; Real Academia Española 1739, 1780). These vessels came in various forms, including wide, bell-shaped, open, and so on. Although archaeologists have suggested that this term also referred to bowls in the sixteenth century (Lister and Lister 1976:83–84; South et al. 1988), eighteenth-century dictionaries offer distinct, unique definitions for *tazas* and *escudillas,* suggesting that the forms and functions were different. One type of cup, identified in eighteenth-century dictionaries and cookbooks as a *xícara* or *jícara*, was noted as being specifically for chocolate consumption. In late eighteenth-century Andalucía, this type of cup was also known as a *pocillo* (Terreros y Pando 1788). To avoid using regional terms that might not be applicable in other areas of the Spanish Empire, this book uses the term *jícara* when referring to chocolate cups. Archaeological examples of *tazas* (or *jícaras*) are small, measuring approximately eight cm tall. 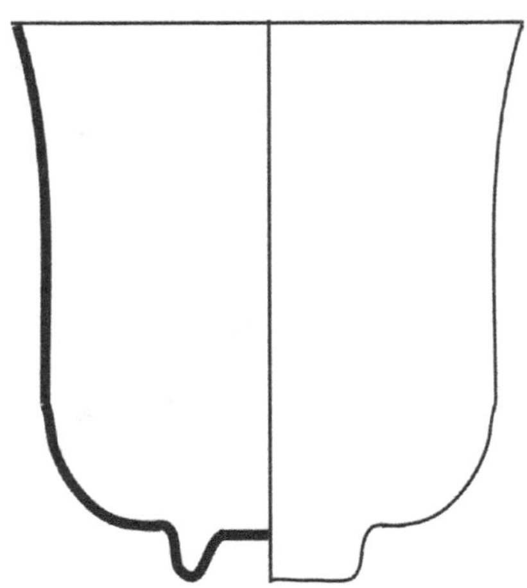

(*continued*)

Ataifor A deep dish for serving meats and stews (Real Academia Española 1726–1739). Late eighteenth-century dictionaries define this form as an outdated term but continue to include it (Real Academia Española 1770). One nearly complete specimen found at La Calle Corredera had a diameter of approximately 34 centimeters. This form is sometimes identified as a *fuente*.

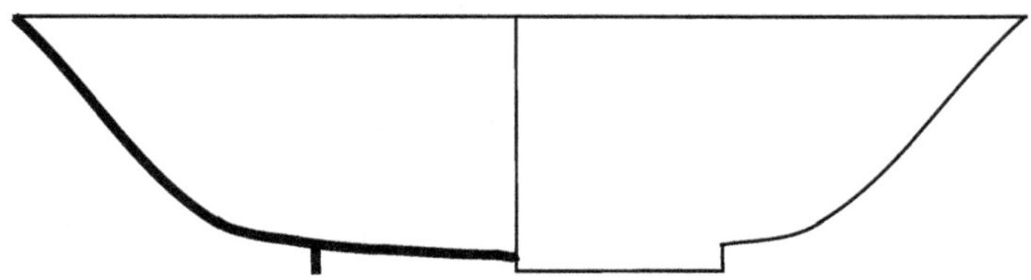

Bandeja A slightly concave tray with a lip around the edge to hold candies, sponge cake, bread, and other foods (Lister and Lister 1976:22–23; Real Academia Española 1770). Eighteenth-century dictionaries define this form as a silver vessel, but archaeologists have recovered ceramic *bandejas* at La Calle Corredera. Similarly, in his analysis of probate inventories, Fernando Carlos Ramos Palencia (2010:191) groups *bandejas* with other ceramic vessels.

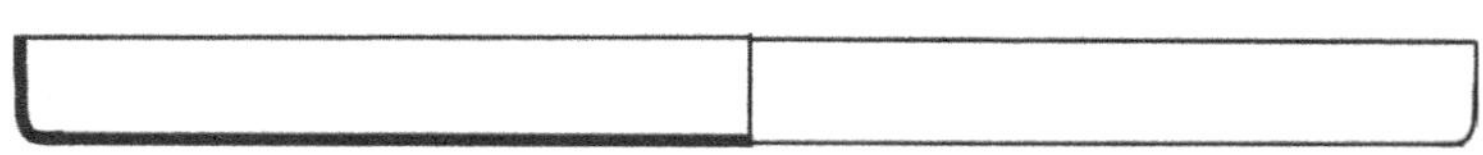

Frutero A small basket or plate designed for displaying and serving fruit (Real Academia Española 1780). While there is no specific shape given for this term, archaeological evidence and ceramic dictionaries indicate that it had a tall, pedestal-like base (Lister and Lister 1976:48; López Rosendo 2005b).

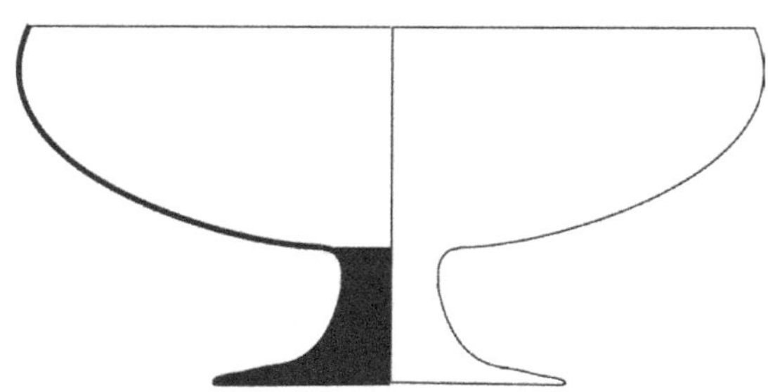

Salero	A salt cellar made of a variety of materials and in a variety of shapes (Lister and Lister 1976:98; Real Academia Española 1780). Archaeological examples from La Calle Corredera are shallow, saucer-like forms that measure approximately eight cm in diameter and have a concave, glazed area in the center for the salt.

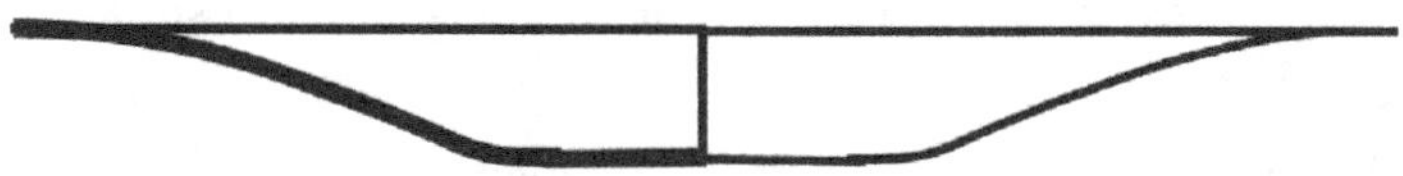

WATER

Jarro	A pitcher with one handle, a wide belly, narrower neck, and a spout for liquid (Lister and Lister 1976:54; Real Academia Española 1780). It is important to note that the same kind of vessel with two handles is known as a *jarra* (Real Academia Española 1780). Because it is often impossible to distinguish between a *jarro* and a *jarra* unless a whole vessel is found, the former term is used in COSA. One semi-complete example from La Calle Corredera measured approximately 18 cm tall.

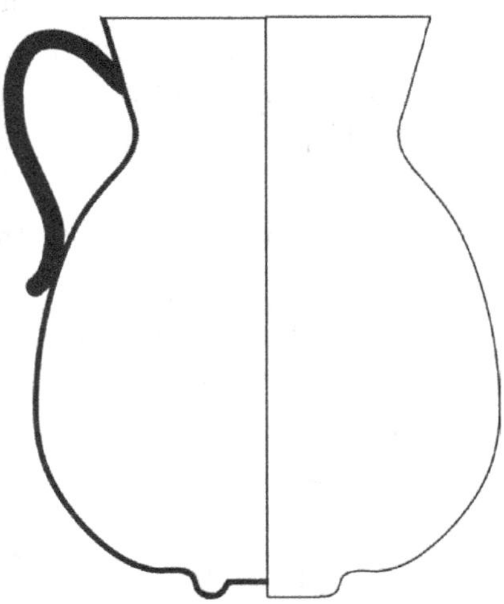

Conclusion

Typologies and classification systems are among an archaeologist's most frequently used tools. They help scholars identify, date, source, organize, and interpret data in a variety of formats. These instruments also provide a standard way to classify artifacts, thereby facilitating intersite comparisons and providing a common terminology for archaeologists working on similar sites.

Noted Caribbean archaeologist Irving Rouse (1960:313) stated, however, that "classification, like statistics, is not an end in itself but a technique by means of which to attain specified objectives, and so must be varied with the objective." A valid archaeological typology or classification system, therefore, is one that can provide insights into past behaviors by allowing scholars to answer a specific set of research questions. In many cases multiple classification systems can and should exist for one group of artifacts, like ceramics, as they facilitate different means of evaluating the data and allow scholars to answer different kinds of questions.

Because the objective of this book is to better understand the Spanish Atlantic identity that developed in the eighteenth century by studying ceramics and dining practices, a classification system was needed that emphasizes the emic perspective and uses of pottery in this region. The vessel-based approach employed in COSA, as well as the inclusion of data from both sides of the Atlantic, makes this classification system ideal for this study.

It is important to emphasize that classification systems like COSA are typically not meant to replace existing typologies. They are different approaches to identifying and organizing the ceramic remains found on sites in the Spanish Atlantic. Both approaches have their strengths and weaknesses, as they help answer different questions and provide distinct ways of looking at and organizing artifacts.

In the case of Spanish ceramics, the existence of multiple classification systems can provide useful insights into different questions regarding the Spanish Atlantic empire. The current Spanish American typology, for example, is invaluable in dating sherds and identifying their manufacturing locations and, by extension, shedding light on issues such as chronology, economics, trade patterns, and cultural exchange. A folk classification system like COSA, in contrast, can yield data regarding vessel forms, daily activities, and cultural perspectives of the users of the ceramics recovered. These systems are not mutually exclusive, as they can and should be used together to obtain a more holistic view of the past. Considering both vessel forms and archaeological ceramic types allows us to better understand the activities and mind-sets of the early modern individuals who constituted the Spanish Empire.

4

La Calle Corredera, Jerez de la Frontera, Spain

Prior to recounting Don Quixote's adventures as a knight-errant, Cervantes gives a brief but informative account about the hero's life as an average gentleman in a village of La Mancha. Three-quarters of Don Quixote's income goes toward food, as he eats a stew with "more beef than mutton" and salad on most nights, scraps on Saturdays, lentils on Fridays, and a young pigeon on Sundays. According to Cervantes, the gentleman spends the rest of his money on his wardrobe, which consists of a broadcloth tunic, velvet breeches and slippers for holy days, and a suit of fine, undyed wool for weekdays. Presumably, some portion of his income also goes toward supporting his housekeeper, a serving boy, and his niece, all of whom live with him, as well as paying for Sancho's expenses (Cervantes 1605 [1964]).

While this description of Don Quixote's daily life is not as entertaining as his many adventures and delusions, it is very telling as to what was important to early modern Spaniards. The fact that the majority of his income went to cuisine and that much of the novel is devoted to a perpetual quest (by Sancho) to find sustenance or professed attempts to ignore such necessities as meals (by Don Quixote) indicates the importance of food to early modern Spaniards. The second-most important expense Cervantes lists is clothing, highlighting the attention many Spaniards paid to their public appearance, a trend that has been observed by archaeologists, historians, and other scholars alike (Bas and Wunder 2009; Donahue 2004; Noyes 1998; Voss 2008b).

Tellingly, the author does not reference other household goods in this description. We do know from later in the novel, however, that Don

Quixote had a very large library of chivalric books and manuscripts and presumably spent significant amounts of money on these texts. Additionally, his household would have included other items, such as ceramics, furniture, and cooking equipment. Because such ubiquitous objects, like pottery vessels, represent a fraction of a household's assemblage and were not as conspicuously ostentatious as fine clothing, they were most likely taken for granted as typical aspects of daily life. As a result, these objects appear to have faded into the background and were ignored, at least by men like Cervantes and Don Quixote.

Despite being overlooked by contemporary users and chroniclers, ceramics represent the most commonly recovered kind of artifact throughout the Spanish Atlantic, frequently representing over half of the archaeological assemblage. These fragments can yield a great deal of information about foodways and activities such as dining, entertaining, and personal hygiene. At domestic sites in Jerez de la Frontera, such as La Calle Corredera, pottery sherds can also indicate how the residents responded to cultural and political changes.

A Brief History of Jerez de la Frontera

Jerez de la Frontera, or Xerez de la Frontera, is an ancient town in southwestern Andalucía, situated approximately 20 km northeast of Cádiz and approximately 60 km south of Sevilla (Figure 4.1). Over the course of its long history, it has been governed by Phoenicians, Romans, and Muslims, before being conquered by Christian king Alfonso X in 1264 (Caro 1999; Castro 1845; Esteve Guerrero 1962). During the Middle Ages, Jerez appears to have been much like other towns in the region, albeit one of the largest ones in the area. By the second quarter of the sixteenth century, Jerez had a population of 3,751 people, more than five times that of Cádiz during the same period (González Beltrán and Pereira Iglesias 1999:27). The population was largely dependent on agriculture, growing and exporting cereals and legumes, and trading fruits, raisins, and wine for products from other parts of Europe (González Beltrán and Pereira Iglesias 1999:65, 67).

In the early eighteenth century, the Casa de Contratación, which handled all legal commercial interactions with the Americas, was moved from Sevilla to nearby Cádiz, as discussed in chapter 2. This change meant that both Cádiz and the surrounding area experienced an economic boom,

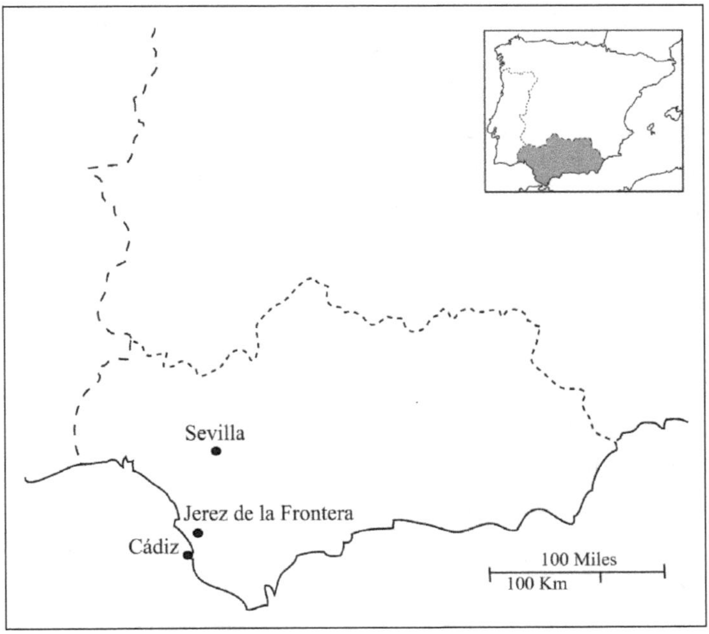

Figure 4.1. Map of Andalucía, Spain. Map drawn by the author.

urban growth, and increased importance. In Jerez the economy expanded to include textile production, leather and metal working, decorative arts, carpentry, construction, and glass-bottle making for the growing wine industry (González Beltrán and Pereira Iglesias 1999:98). Despite such diversification, these developing industries employed only approximately 17 percent of the population, and viniculture continued to be the staple of the local economy, with sherry being the town's most important product (González Beltrán and Pereira Iglesias 1999:98). Jerez became so well known for its wine that it was frequently mentioned by foreign visitors to Andalucía (Dennis 1839:132–133; Dillon 1781:373, 466; Roscoe 1836:220–226).

During this period, the population grew significantly, and the city experienced urban renewal and expanded beyond the walls of the medieval town (Aroca Vicenti 2002; Esteve Guerrero 1962:17; López Rosendo 2005c:34). While the improved economic situation was attractive, many people also moved to Jerez after the 1755 earthquake in Lisbon caused a tsunami that flooded coastal towns like Cádiz and nearby El Puerto de Santa Maria (López Rosendo 2005c:34). By 1787 the town had swelled to 45,506 residents (González Beltrán and Pereira Iglesias 1999:37).

Many of the families who settled in the area became involved in the wine industry and set up bodegas to produce sherry on a large scale. With the proceeds from these successful endeavors, they built grand houses and contributed to the town's expansion (Aroca Vicenti 2002:142–143; Esteve Guerrero 1962:17). Interestingly, these bodegas are often located within Jerez, with one bodega less than a kilometer from La Calle Corredera and the center of town.

La Calle Corredera is situated in one of the neighborhoods that grew during this period, San Miguel. This parish had, and still retains, distinct characteristics that separate it from the medieval portion of town and other sections of town developed in the eighteenth century (Aroca Vicenti 2002:119). Located east of Jerez's medieval walls, this area is the largest of the eighteenth-century neighborhoods, featuring a mixture of long, wide boulevards as well as smaller, narrower streets and plazas. San Miguel's three major streets, Arcos, Medina, and Corredera, all originated at various points along the town's medieval walls and expanded outward. Corredera is particularly noteworthy in this list, as it ran along the Puerta Real and was the most direct route to agricultural fields and pastures outside the urban area (López Rosendo 2005c:3–4).

By 1775 San Miguel had 111 streets and 1,697 houses (Aroca Vicenti 2002:104, 127). While the town's population declined slightly in the fourth quarter of the eighteenth century, this neighborhood continued to be the most populous, with the 1787 census listing 755 houses and 16,090 residents (Aroca Vicenti 2002:103). In addition to its significant residential population, San Miguel was home to a Franciscan convent, San Francisco, and its related church, both of which shared property lines with the site La Calle Corredera. One of the main attractions in the neighborhood, however, was the San Miguel church, which was mentioned in at least one nineteenth-century guidebook as a key destination for any visitor to the town (Roscoe 1836:226).

By the 1800s Jerez was a well-known stopping point for British visitors on their way to Sevilla or Cádiz, prompting many to write interesting and often amusing accounts of the town during this period. While some praised the town's vistas, architecture, and products (Roscoe 1836:217–219), others argued that it was not as pretty or clean as Cádiz (Dennis 1839:135–136). Most of the accounts concur that, while the area had lost some of its splendor by the nineteenth century and had a reputation for being somewhat dangerous, it still had many attractions worth visiting.

Jerez continues to be highly regarded among tourists, who stop there to visit its numerous historical and cultural highlights. Today the top attractions include the Alcazar, or Moorish fortress, the many bodegas producing world-famous sherry, the cathedral, the archaeology museum, and the Royal Equestrian School, or Fundación Real Escuela Andaluza del Arte Ecuestre. Visitors journey to the city for a variety of reasons, some combining trips to the town with stays in Sevilla or the beaches of the Costa de la Luz, and others staying for longer periods to study Spanish guitar or flamenco.

Archaeology in Jerez de la Frontera

Although many of Jerez's modern attractions are not archaeological, the town has a strong archaeological tradition, like other areas of Andalucía. By the nineteenth century, locals had created an archaeological museum in the Plaza de la Asunción to publicly display artifacts found throughout the city (González Rodríguez et al. 1997:12–18). Archaeological research continued, and in the 1990s the museum was moved to the Plaza de Mercado in the north of the city (González Rodríguez et al. 1997:16–18), where it remains today.

Since 1986 many local excavations have been published in the *Anuario arqueologico de Andalucía*, an annual publication of the Junta de Andalucía that documents the wide variety of archaeological research conducted in the region. The majority of these projects, including La Calle Corredera, are public works or cultural resource management activities (Andalucía Dirección General de Bienes Culturales 1985–2006). In 2008 officials in Jerez summarized the various kinds of protected sites and archaeological zones in Jerez, excavations done to date, and procedures for future excavations (González Rodríguez et al. 2008). While the majority of these projects focused on pre-*moderna* era sites, there have been a number of studies of early modern sites and buildings, and this period is recognized as an integral part of the town's history (González Rodríguez et al. 2008).

Archaeology at La Calle Corredera

La Calle Corredera is one of the few published excavations of a *moderna* era site from this area. While this site is located just outside the medieval town walls, it was on a major thoroughfare in the San Miguel parish that, as mentioned above, experienced substantial growth and revitalization during the eighteenth century. Although we do not have the names of the eighteenth-century residents, excavations at the site have revealed a wealth of information, both about early modern daily life in this burgeoning, upper-middle-class neighborhood and about the development of the site itself.

In October and November of 2005, Spanish archaeologist Ester López Rosendo conducted the first archaeological excavation on this street at La Calle Corredera, 39 (López Rosendo 2005c:4). The project was prompted by new construction that involved tearing down the existing building and erecting a new residential structure with an underground parking garage (López Rosendo 2005c). As with other archaeological activities in the city, López Rosendo's work was an attempt to document and preserve any cultural heritage at the site prior to excavating the garage and constructing the new building (López Rosendo 2005c:4).

According to her excavations, the site had numerous building phases that occurred from the sixteenth to the nineteenth century. Indeed, the structure that was demolished in 2005 was most likely built sometime in the second half of the eighteenth century; as López Rosendo notes, it had been heavily modified in the nineteenth century and was listed in nineteenth-century town records as a bourgeois house (López Rosendo 2005c). This structure would have been a typical, multistory Spanish house, with a central courtyard that had rooms arrayed around it, and possibly a smaller courtyard near the rear of the lot where cooking activities might have taken place.

López Rosendo conducted systematic excavations of three areas of the site and uncovered portions of the family's living areas, central courtyard, stables, and a back patio. It is perhaps worth noting that, while the excavation covered only the lot at La Calle Corredera, 39, the original property also included what is now La Calle Corredera, 41. Number 41, however, was not excavated; the data here are exclusively from number 39 and therefore represent only the western half of the original site.

Overall, lot number 39 contained nearly 6,700 artifacts, over 5,000 of which were ceramic sherds, ranging in date from the Islamic period to the twentieth century. In the site map (Figure 4.2), shaded triangles mark the location of the two courtyards, one near excavation unit C-1 and another near excavation unit C-3, while the floor plan depicts the structure that

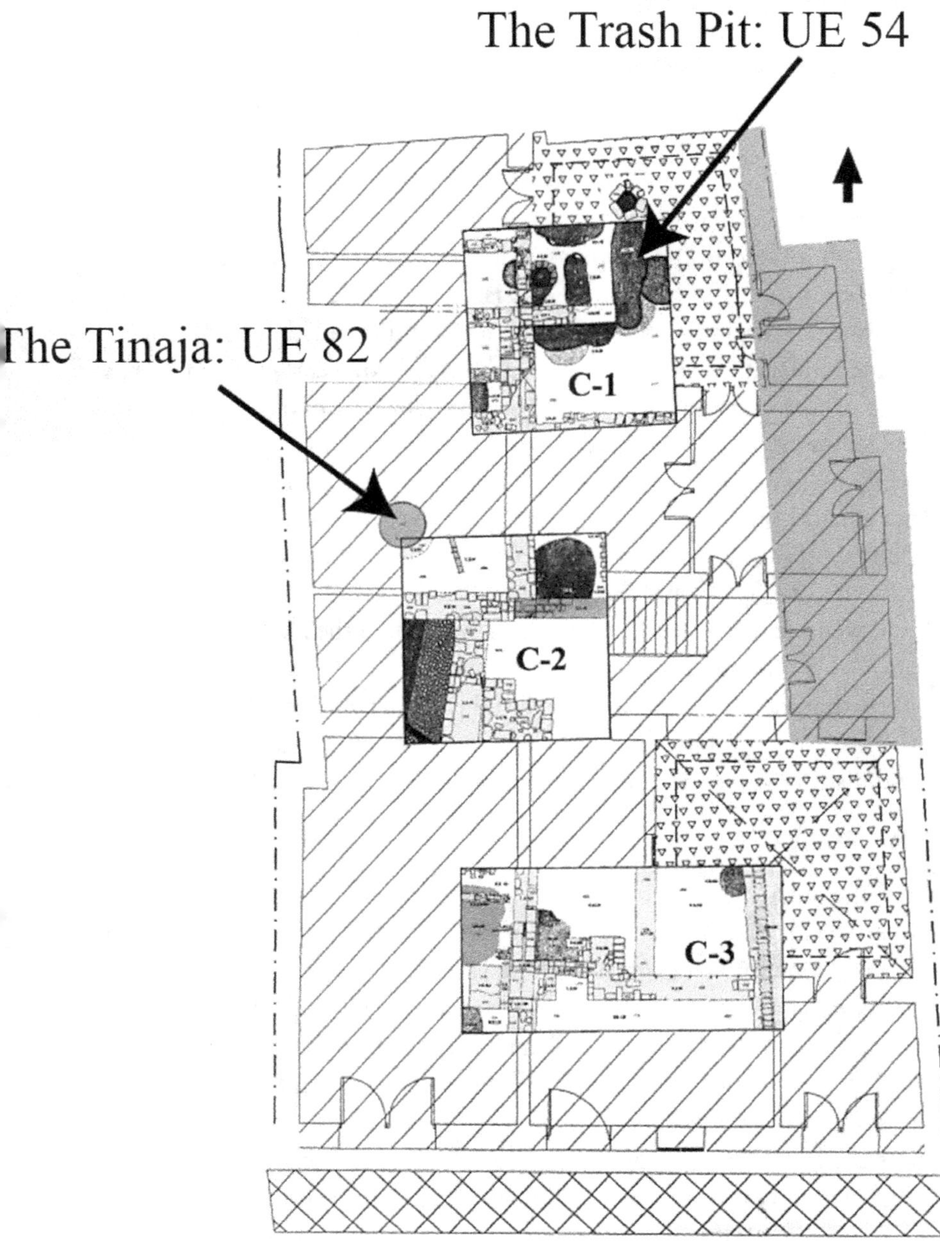

Figure 4.2. Map of the excavations at La Calle Corredera with the street at the bottom of the plan; C-1, C-2, and C-3 represent excavation units. Image adapted from López Rosendo 2005c:52.

was demolished in 2005. Although the majority of eighteenth-century features corresponded to building activities from the second half of the century, there were two features that related to earlier domestic activities—UE 54 and UE 82—both of which are the basis for this study.

The Trash Pit: UE 54

The feature known as UE 54 was a seventeenth-to-eighteenth-century trash pit located in the northeast portion of the first unit excavated, C-1. The ceramics from this feature consist of a mixture of utilitarian and tableware sherds, many of which were large and mended to form whole or semicomplete vessels. While this trash pit included earlier ceramic types, such as Columbia Plain (1490–1650; Florida Museum of Natural History 2012), many of the tableware sherds appear to be finer quality, blue-on-white majolica that are similar to those made in Triana, Sevilla, around the second third of the eighteenth century (Pleguezuelo 2011). A small number of sherds were part of the Bérain series of blue-on-white ceramics from Alcora, Spain, and date between 1727 and 1749.

Given the location of the feature at the back of the house and the presence of numerous utilitarian cooking vessels and faunal remains, it is likely that the trash pit was related to the kitchen and that the material in the trash pit combines both kitchen trash accumulated during cooking and dishes that were stored, used, or broken in that area of the house.

The *Tinaja*: UE 82

On the northwest corner of the second excavation unit, C-2, López Rosendo (2005c:9) found a large *tinaja*, designated UE 82, that was installed sometime in the sixteenth or early seventeenth century, possibly to serve as a cistern. Because this location would have been the main living area of the house, the fill most likely represents material from a variety of activities. The ceramics from this feature largely resemble and, in some cases, mend to sherds from the trash pit, indicating that the jar was used as a trash receptacle in the mid-eighteenth century prior to the remodeling that took place in the late eighteenth century. Additionally, the presence of one Whieldon-ware bowl base sherd (1740–1770; Florida Museum of Natural History 2012) indicates that the *tinaja* was used as a trash pit around the 1740s.

Examination of Artifacts

In both features the ceramics were fairly large and included a significant number of crossmends, suggesting that they were deposited at the same time and most likely correspond to one family moving or cleaning out its household prior to remodeling or constructing a new building. If so, then these pits provide valuable insight into one family's possessions and their desire to remain fashionable by disposing of older tableware vessels.

Ceramics from the Trash Pit and the *Tinaja*

The ceramics from this site exhibit a wide variety of decorative techniques and vessel forms. Between the trash pit and the *tinaja*, the ceramic assemblage includes at least one vessel in nearly every category represented in COSA (Figures 4.3 and 4.4; Ness 2015a). When sixteenth-to-seventeenth-century features at the site are compared to the eighteenth-century features discussed above, it becomes apparent that there was a shift in both the vessel forms and aesthetics of tableware. These changes appear to be in keeping with the rise in French fashions in Spain and indicate that individuals living in this area of Jerez were attempting to follow court trends from Madrid.

The tableware from older features at the site, for example, consists almost exclusively of *platos viejos* and *escudillas viejas*. The relative lack of *tazas* or other beverage vessels in these earlier features suggests that the occupants might have drunk from *escudillas* or straight from *jarros*, despite this use not being documented in early modern texts for either vessel. All the *platos viejos* and *escudillas viejas* recovered throughout the site are thick-bodied and feature a white, chalky paste and white or off-white glaze, making them analogous to a type known in the Americas as Columbia Plain and in Spain as Loza Blanca.

While nine Columbia Plain *platos nuevos* and three Colmbia Plain *escudillas nuevas* were found in the two features, the majority of these forms were typically made of blue-on-white majolica. Unfortunately, it is currently impossible to identify definitively the origins and precise manufacturing dates of these ceramics, because of the lack of a detailed typology for majolicas within Spain. Additionally, many of these ceramic types made in Spain were not commonly imported to the Americas and therefore are not identified in the Spanish American typology. The majolica

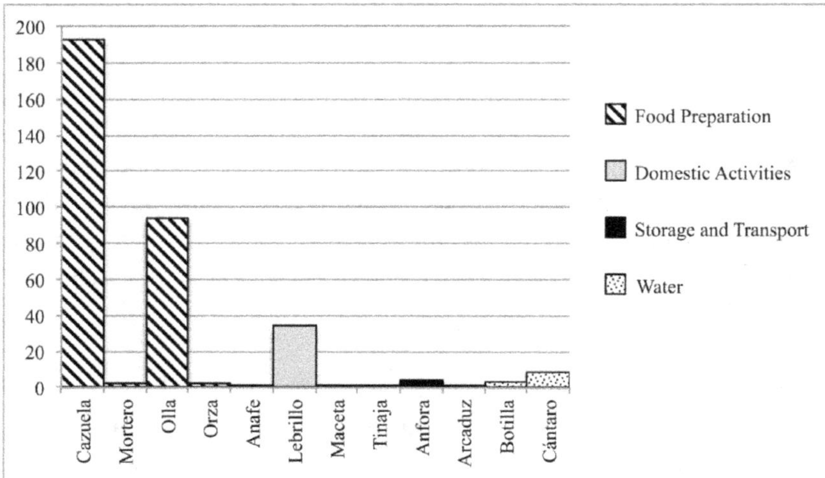

Figure 4.3. Minimum number of vessels (MNV) of *loza común* from UE 54 and UE 82.

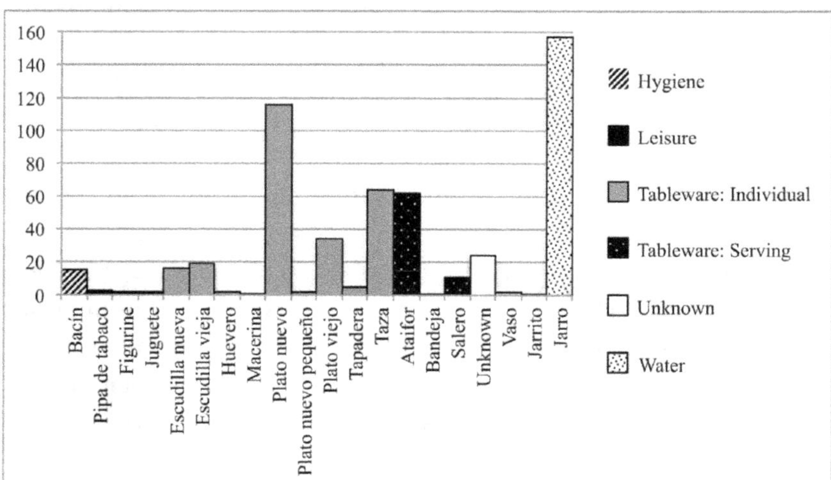

Figure 4.4. Minimum number of vessels (MNV) of *loza fina* from UE 54 and UE 82.

types found at La Calle Corredera that are mentioned in the Spanish American typology typically fall under the Spanish American category Unidentified Blue on White, Iberian.

The difficulty in identifying these sherds is apparent in the fact that López Rosendo (pers. comm. 2013) classifies them as being from the Cádiz Bay area, while Ruiz Gil (pers. comm. 2013), a local ceramicist and archaeologist, argues that the same vessels are from the Netherlands.

American scholars, seeing pictures of these vessels, agree with Ruiz Gil and classify them as delftware (G. Waters, pers. comm. 2013). However, based on the similarity of many of these sherds to ceramics produced in Triana, Sevilla, during the middle of the eighteenth century (Pleguezuelo 2011), it seems likely that the majority of blue-on-white vessels from La Calle Corredera were manufactured in Triana, Sevilla, during the second third of the eighteenth century.

Matching Dishes

While the concept of sets was popularized with mass production and the advent of well-known businesses such as Wedgwood in the middle of the eighteenth century, matching objects were available to consumers as early as the seventeenth century in the form of upholstered furniture (Hellman 2007:130; Thornton 1978:103–106). By the early eighteenth century, this concept had spread to other aspects of material culture, including metalwork, decorations, and ceramics (Hellman 2007:130). The predilection for matching items spread throughout Europe; however, it was particularly prominent in France (Hellman 2007:130).

The archaeological evidence from La Calle Corredera suggests that Spaniards had also embraced this love of aesthetic repetition, possibly as part of their attempts to imitate French fashion. Between the two features from this site, there were several matching eighteenth-century *platos nuevos* and *escudillas nuevas,* and at least one possible set of two *platos nuevos* and two *escudillas nuevas* (Figure 4.5). These vessels appear to have been made at the same factory in Triana, which suggests that they were produced and possibly purchased as a set.

The assemblage at La Calle Corredera also contains examples of at least three distinct and elaborate patterns that match those on the vessels from Triana, Sevilla, that are part of the Colección Carranza at the Real Alcázar in Sevilla (Figures 4.6, 4.7, and 4.8). Both the similarities between the two collections and the fact that so many vessels with the same patterns were found at La Calle Corredera suggest that the motifs were common decorations from potteries in Triana, Sevilla, and might have been popular decorations for sets of dishes produced in the area.

Other vessels from this site, in contrast, have matching decorations but seem to have been made in different locations. At least one *taza* was clearly made at the Real Fábrica de Alcora near Valencia and has the lacy Bérain decorations, while one *escudilla nueva* and one *plato nuevo* were

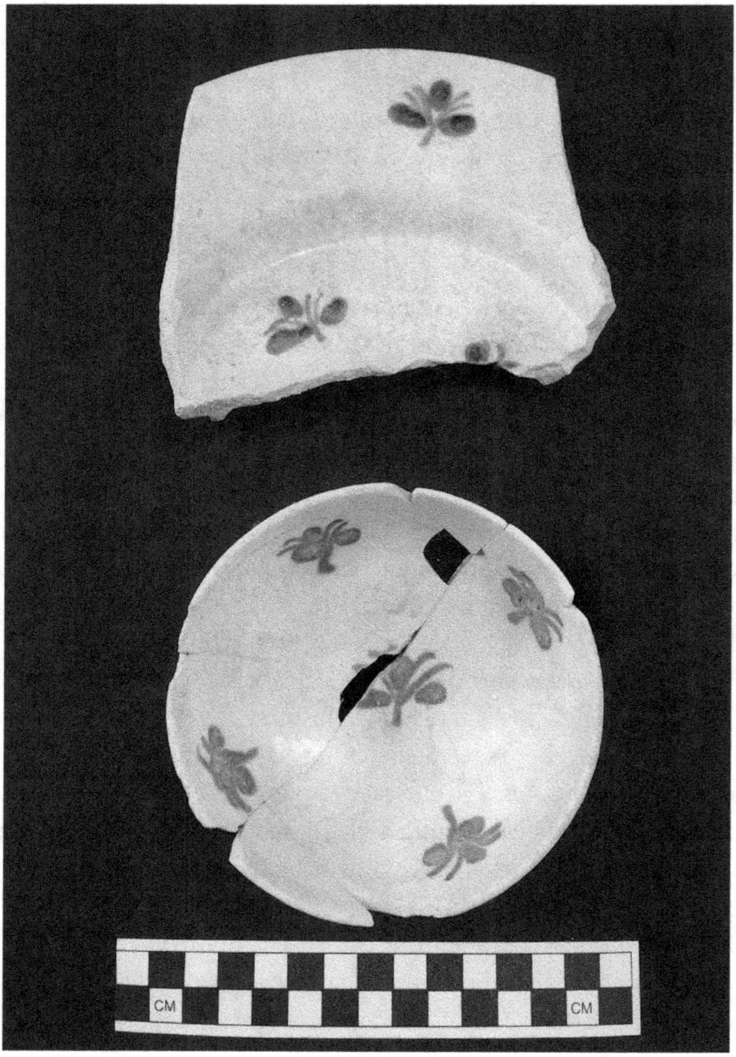

Figure 4.5. Part of a set of matching blue-on-white majolica *platos nuevos* and *escudillas nuevas* from the trash pit (UE 54), deposited in the Museo Arqueológico Municipal de Jerez de la Frontera. Image courtesy of the Museo Arqueológico Municipal de Jerez de la Frontera and the Junta de Andalucía.

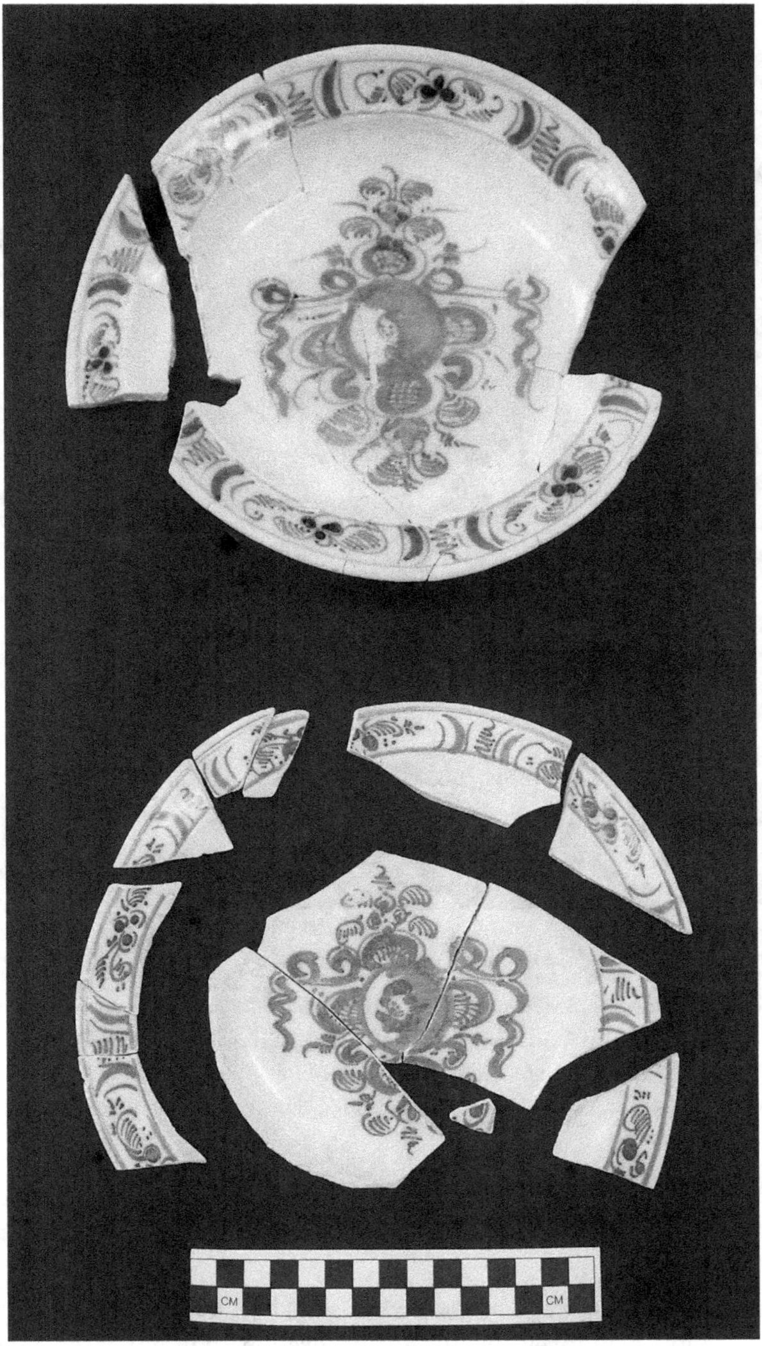

Figure 4.6. Matching *platos nuevos* most likely made in Triana and recovered from the trash pit (UE 54), deposited in the Museo Arqueológico Municipal de Jerez de la Frontera. Image courtesy of the Museo Arqueológico Municipal de Jerez de la Frontera and the Junta de Andalucía.

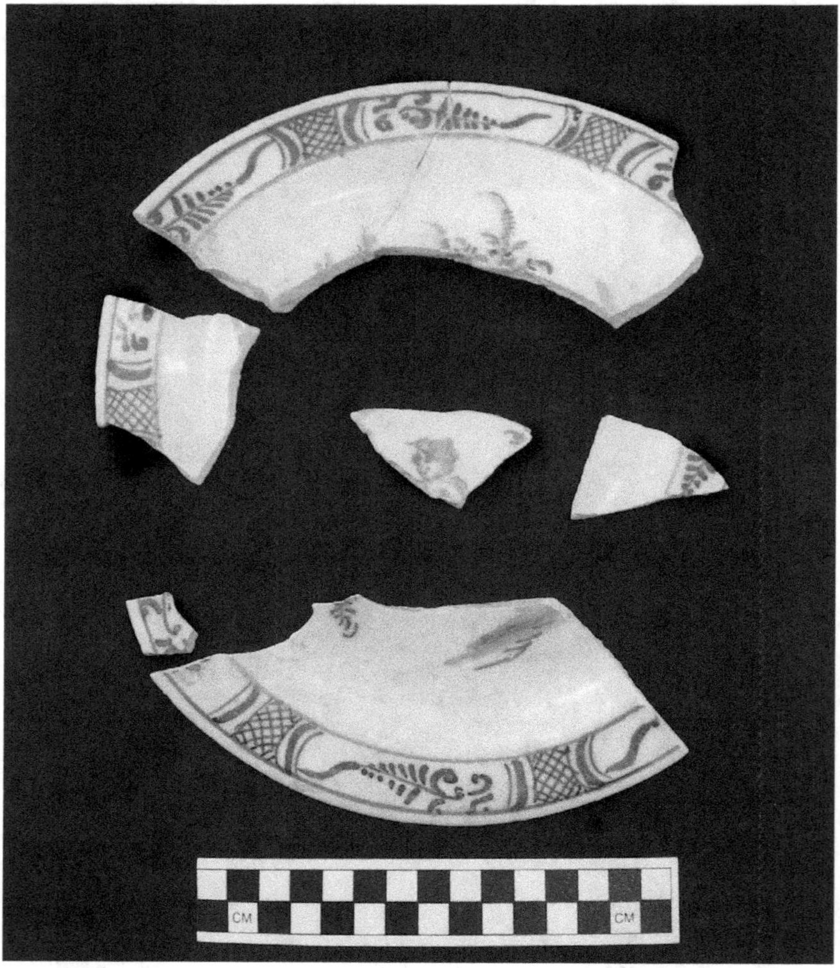

Figure 4.7. A *plato nuevo* most likely made in Triana and recovered from the trash pit (UE 54), deposited in the Museo Arqueológico Municipal de Jerez de la Frontera. Image courtesy of the Museo Arqueológico Municipal de Jerez de la Frontera and the Junta de Andalucía.

each made elsewhere, possibly Triana, Sevilla, but have the same Bérain-style motif (Figure 4.9).

Although the fact that one factory was copying another makes it difficult to identify definitively manufacturing locations, it indicates that there were popular motifs that were imitated throughout Spain. Furthermore, this copying suggests that consumers might have mixed and matched sets, purchasing vessels from different factories either to obtain a complete set

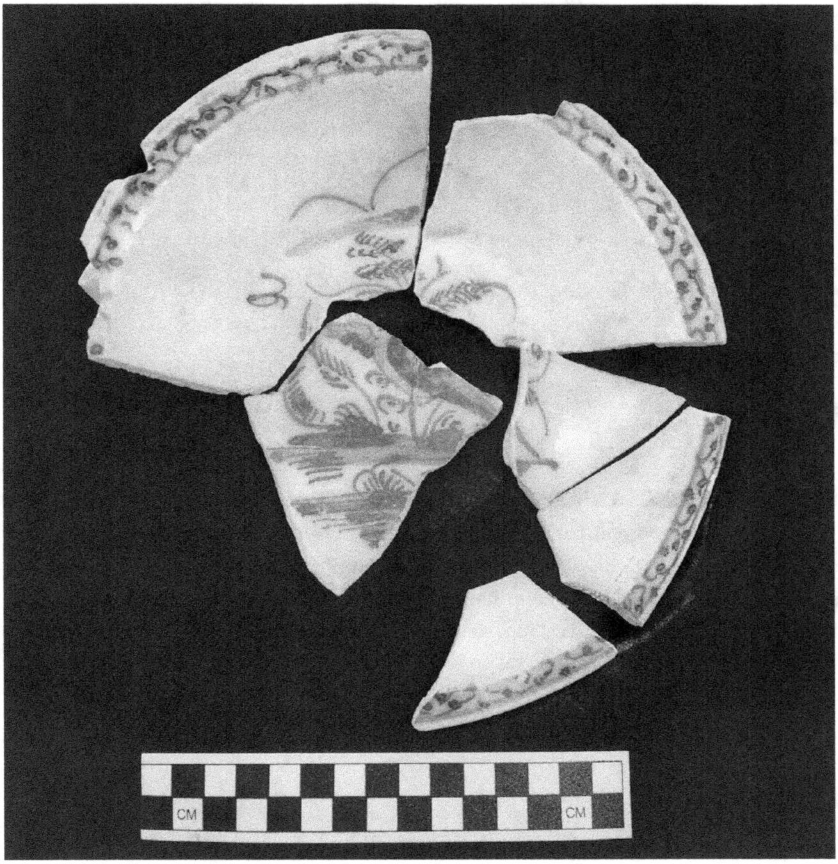

Figure 4.8. An *ataifor* most likely made in Triana and recovered from the trash pit (UE 54), deposited in the Museo Arqueológico Municipal de Jerez de la Frontera. Image courtesy of the Museo Arqueológico Municipal de Jerez de la Frontera and the Junta de Andalucía.

at a more economical price or in an attempt to replace broken vessels to extend the life of a set.

While elemental analyses have yet to be performed on these kinds of ceramics to determine the precise manufacturing location and possible means of assembling sets, numerous studies have demonstrated that it is possible to identify specific production centers in both Spain and Mexico based on the elemental composition of the clay (Garcia Iñañez et al. 2008, 2009a, 2009b; Myers et al. 1992; Olin et al. 1978; Olin and Blackman 1989). Future studies might prove useful in further examining this question.

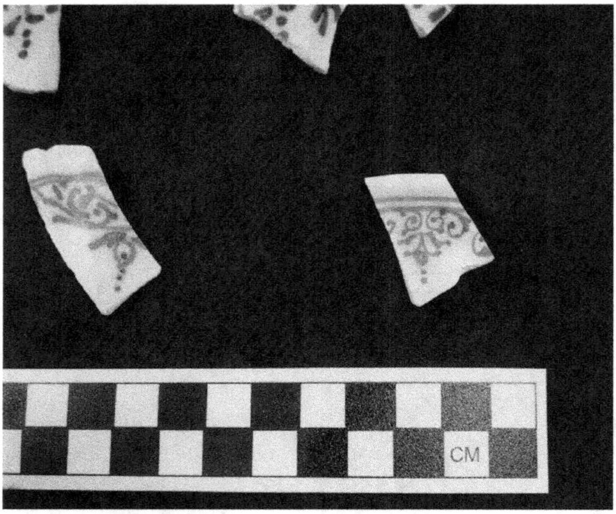

Figure 4.9. Rim fragments from two *jícaras* found in the trash pit (UE 54) and the *tinaja* (UE 82), deposited in the Museo Arqueológico Municipal de Jerez de la Frontera. The sherd on the left is most likely an imitation of the delicate Bérain style ceramic produced in Alcora around 1727, which is represented by the sherd on the right. Image courtesy of the Museo Arqueológico Municipal de Jerez de la Frontera and the Junta de Andalucía.

Tableware

The presence of sets is especially noticeable, given the substantial amount of tableware found at the site. Between the trash pit and the *tinaja*, nearly 60 percent of the vessels recovered were tableware (Ness 2015b). While the tableware category did include some serving dishes and water vessels, which will be discussed below, nearly half of the vessels in this category were intended for individual use. These deposits contained a surprising 116 *platos nuevos* and only 16 *escudillas nuevas*. This significant disparity suggests a shift away from traditional stews and broth-based dishes in favor of foods that could be consumed on flatter plates.

In addition to adopting new foods and cooking techniques, eighteenth-century *jerezanos* appear to have developed a taste for chocolate, as indicated by the 64 *tazas*, or *jícaras*, and one *macerina* that were recovered in these deposits but not in earlier features. These cups, which were the second-most common individual tableware vessel after *platos nuevos*, speak to changes in both the purchasing power and diet of *jerezanos* in the

San Miguel parish during the eighteenth century and indicate that these individuals were able to afford and consume imported foods such as chocolate. The *tazas* appear to originate from several different manufacturing locations, with at least one from the Real Fábrica de Alcora. Additionally, one of the only polychrome vessels from the eighteenth-century features was a *taza*, suggesting that while locals may have enjoyed the idea of a matching set, it might not have been seen as a requirement.

Although the *tazas* were the only individual ceramic beverage vessels in these features, the deposits yielded 157 *jarros*, tableware vessels for water and possibly other beverages. Four of the *jarros* were made of a ceramic type known in the United States as Santa Elena Blue on White (South et al. 1988), and in Spain as *azul motado*, a majolica with mottled blue decorations on a white background. The majority of these jars, however, were made of a type known as *paredes finas* (Figure 4.10). This unglazed, thin-walled, fine, buff earthenware has been interpreted as an early modern copy of both Roman and medieval pottery (López Rosendo, pers. comm. 2013), making it difficult to date precisely. The delicacy of these vessels suggests that they might have been subject to high breakage, which would account for their frequency in these two features. Despite their potential tendency to break, these vessels appear to have been frequently replaced, indicating the importance of having water available at the table. Additionally, the fact that they continued to use *paredas finas* as opposed to majolica to make *jarros* suggests that this ceramic type was important for this particular form.

The third most common tableware vessel recovered from the trash pit and the *tinaja* is the *ataifor*, or deep dish. While *ataifores* consisted of only seven percent of the combined assemblage from the two features, those present were primarily in the trash pit. With one polychrome exception, these dishes were blue-on-white majolica but not of a type that matched the *platos nuevos, escudillas nuevas*, or *tazas*.

The relative scarcity of *ataifores* at this site could be accounted for in a variety of ways. This term and form was introduced in the Middle Ages by the Muslims (Real Academia Española 1726–1739) and therefore might have been considered an old-fashioned vessel that was rarely used by the eighteenth century. In terms of use, one household would need fewer *ataifores* than *platos* or *escudillas*, because each individual diner would have his or her own place setting, but the hosts would need only one serving

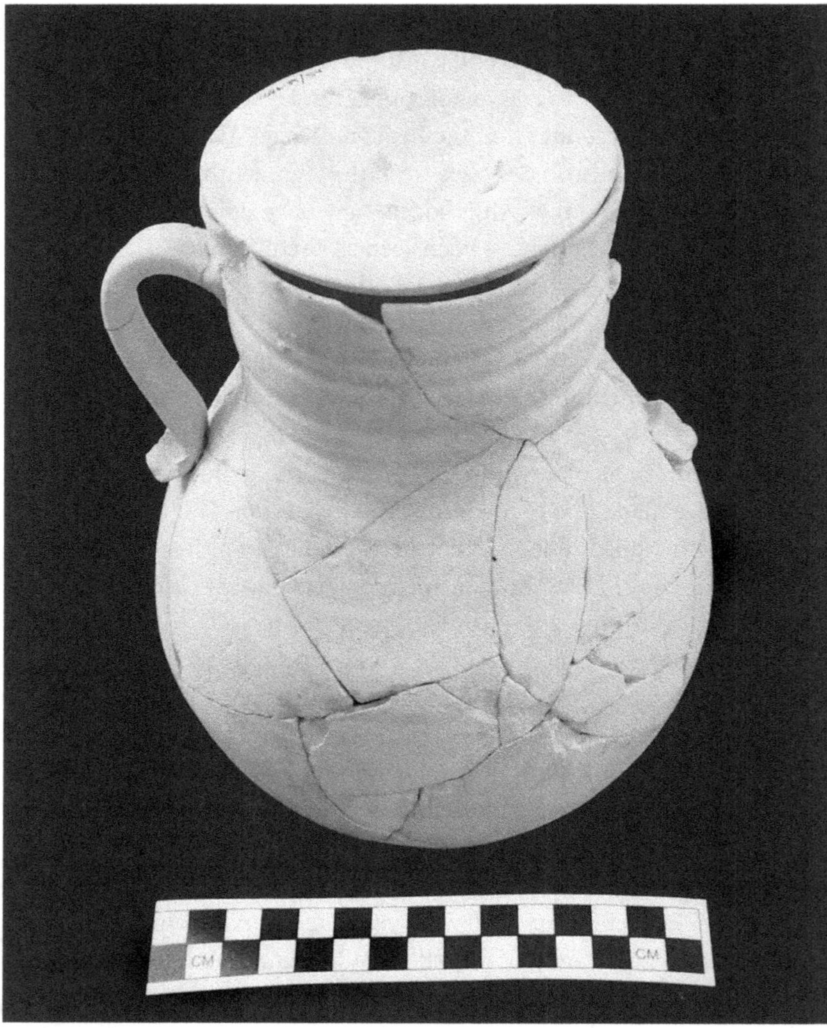

Figure 4.10. A *paredes finas jarro* and *tapadera* from the trash pit (UE 54) and the *tinaja* (UE 82), deposited in the Museo Arqueológico Municipal de Jerez de la Frontera. Image courtesy of the Museo Arqueológico Municipal de Jerez de la Frontera and the Junta de Andalucía.

vessel per culinary dish. Consequently, there should be fewer *ataifores* in the archaeological assemblage, because there would be fewer in the household. Additionally, *ataifores* might have been passed to descendants as heirlooms more often than other dishes, as this form seems to be among the most common to appear in museums with early modern ceramic displays.

Utilitarian Dishes

Just as the tableware fragments provide insight into changes at the eighteenth-century *jerezana* dining table, utilitarian vessels also yield information about dining practices. The main cooking vessels recovered in the trash pit and the *tinaja* were *cazuelas* and *ollas*. These were exclusively made from a coarse red paste and lead-glazed interior with an unglazed exterior (Figures 4.11 and 4.12), and many had scorch marks on the exteriors, indicating that they had been used, some fairly heavily.

While earlier features from this site also included *cazuelas* and *ollas* that were similar to those found in the eighteenth-century deposits, some of these earlier features were disturbed, making it difficult to determine

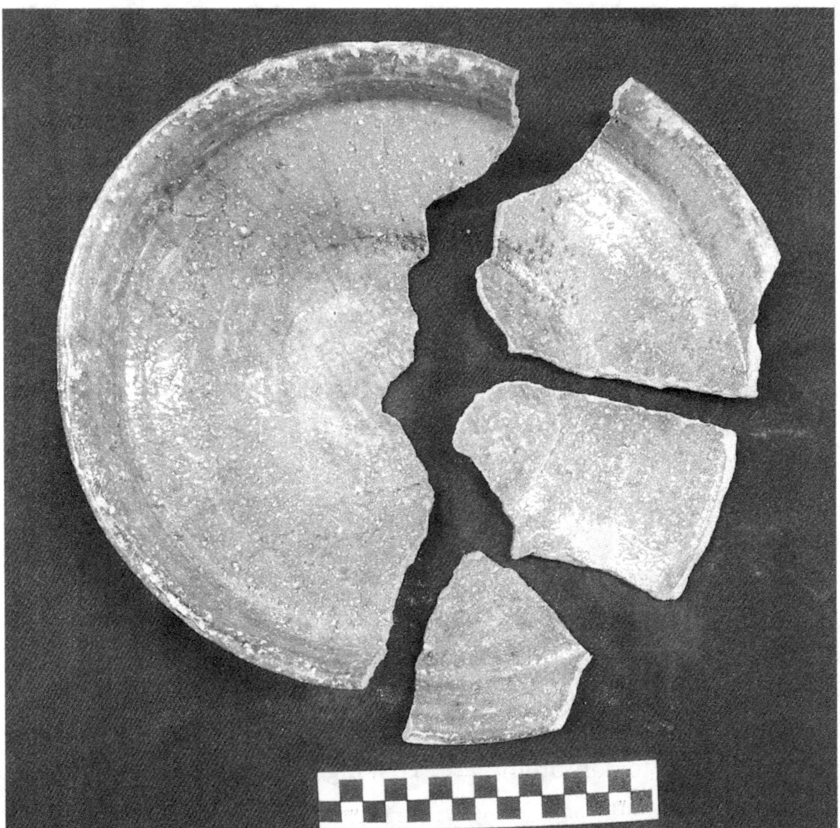

Figure 4.11. *Cazuela* fragments from the trash pit (UE 54), deposited in the Museo Arqueológico Municipal de Jerez de la Frontera. Image courtesy of the Museo Arqueológico Municipal de Jerez de la Frontera and the Junta de Andalucía.

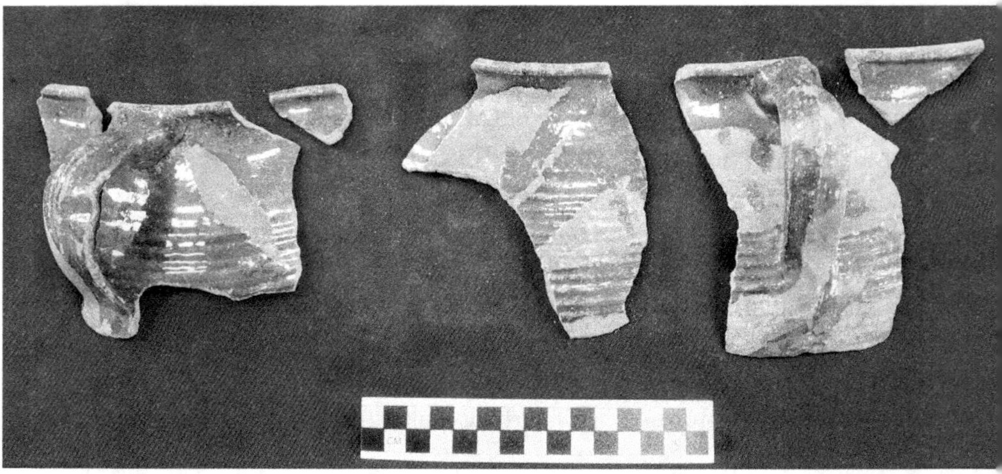

Figure 4.12. Fragments from two *ollas* found at the trash pit (UE 54), deposited in the Museo Arqueológico Municipal de Jerez de la Frontera. Image courtesy of the Museo Arqueológico Municipal de Jerez de la Frontera and the Junta de Andalucía.

when these forms first arrived at the site. There is evidence, however, to suggest that these forms were in use prior to the eighteenth century, because they have been found at medieval sites and appear in seventeenth-century paintings such as *Old Woman Frying Eggs* by Diego Velázquez. Additionally, the terms *cazuela*, its variant *caçuela*, and *olla* appear in sixteenth- and seventeenth-century Spanish dictionaries and vocabularies (Minsheu 1617; Percival 1591), and, as early as 1611 (Covarrubias 1611) dictionaries used the term *olla* to indicate a stew cooked inside the pot.

Cazuelas and *ollas* were among the most commonly recovered forms, representing nearly 83 percent of the *loza común* ceramics and approximately 40 percent of the overall ceramic assemblage from the two features in question. The changing cooking methods are perhaps indicated by the disparate numbers of these vessels, because the two deposits produced nearly twice as many *cazuelas* as *ollas*. While the latter form was typically used to make stews, *cazuelas* are an open form that would have been more conducive to non-broth-based cooking techniques, such as browning meats or sautéing.

Just as individuals continued to make and use the same shape for *cazuelas* and *ollas*, *lebrillos*, which could have served utilitarian, hygiene, food preparation, or serving purposes, also appear to have been made in the same shape over this period. This multifunctional form was the third-most commonly recovered *loza común* vessel in the eighteenth-century deposits, constituting nearly 10 percent of this category. This form includes perhaps the widest variety of decorations of any of the vessel forms recovered at La Calle Corredera. While all have the typical, wide rectangular rim and shallow body, the sherds indicate that these vessels could have been unglazed, incised, glazed with a green lead glaze, decorated with only stripes or wavy lines of the green lead glaze on a white slip background, or decorated with a thin, blue-on-white glaze on the interior. Unfortunately, it is not possible to determine from the assemblage at this site precisely how these vessels were used and if the decorations relate to different uses.

Conclusion

This site is invaluable because it is one of the few early modern domestic sites from this area to have been excavated and published. The artifacts and their distribution throughout the trash pit and the *tinaja* suggest that the objects belonged to one family who occupied the site during the first half of the eighteenth century and deposited old or broken vessels either prior to moving or prior to building a new house on the same site. The relatively high concentration of utilitarian ceramics and faunal remains in the trash pit, at the rear of the lot, suggests that the kitchen was located in this area and that the deposit might have been connected to this room.

While the combination of sixteenth-, seventeenth-, and eighteenth-century features makes the site ideal for examining changes in domestic activities and material culture over time, the wealth of eighteenth-century ceramics also facilitates research on this period. The high concentration of both *platos nuevos*, including some matching sets, and *cazuelas* suggests that the residents were transitioning away from traditional broth-based foods, like the stew colloquially known as an *olla*, in favor of new, relatively drier foods and cooking techniques. This new cuisine was accompanied by water or other beverages that was served in and possibly consumed from the many *jarros* recovered at the site. Additionally, the presence of *tazas* and a *macerina* for chocolate indicates that those living at La Calle Corredera would have had access to imported foods as well as the specialized vessels associated with their consumption.

Although early modern sites with excavated and documented eighteenth-century features like La Calle Corredera are rare in Spain, they are relatively common in Spanish America, particularly in Florida and the Caribbean. In the next chapter, I examine two such households from St. Augustine, Florida, and how their residents responded to changes such as those reflected in the eighteenth-century deposits at La Calle Corredera.

5

The Ponce de León and de Salas Households, St. Augustine, Florida

Early in his adventures, Don Quixote meets a caravan of travelers on their way to Sevilla. Among this group of people is a woman from Biscay, on Spain's north coast, who is traveling to Sevilla to meet her husband and board a ship for the Indies. The woman's party consists of four or five people on horseback and two friars who have temporarily joined them. Upon seeing the party, Don Quixote becomes convinced that the woman is a kidnapped princess and that it is his duty to free her from her "captors." The Knight of the Rueful Countenance attacks the friars and is about to kill her squire as the final act of "freeing" the woman when Cervantes abruptly ends the tale, explaining that he was unable to find more information about this particular adventure than what he has already shared (Cervantes 1605 [1964]).

This particular tale, although brief, is fitting for an analysis of eighteenth-century St. Augustine in several respects. Cervantes provides details regarding the battle, the number of people traveling, and their vehicles and animals (a coach, four or five horses, and two dromedaries), but he says nothing about what the lady is bringing with her to her new home in the Americas. Since her husband had "an appointment of high honor," one could assume she would have been wealthy enough to bring some personal possessions. Unfortunately for those interested in material culture, Cervantes ignores her possessions and leaves it to the reader to imagine what objects were with her in the coach.

Similarly, writers of more official texts often paid little to no attention to what individuals brought with them to the Americas or what items they deemed precious enough to carry themselves, as opposed to objects that could be repurchased in their new cities. While household inventories from this era are often fairly detailed and useful, they do not always indicate what objects individuals would have taken with them when moving to a new continent.

In the case of eighteenth-century St. Augustine, the entire town was evacuated in 1763 when Florida was ceded to the British government for 21 years. Archaeologists studying deposits from this period typically work with the goods that were intentionally left behind and dumped in pits or wells. While it is sometimes difficult to fill in the gaps for what exactly the families might have taken with them or passed along as heirlooms, the discarded items can be particularly telling indicators of what was considered replaceable.

In addition to omitting detail regarding the material culture, Cervantes abruptly ends his tale with the excuse that he could not find any more information on how the adventure ended. Because this incident happens near the beginning of the book, it is clear that Don Quixote survived relatively unscathed, although the lady's squire might not have been so fortunate. While Cervantes's audience might have felt that this adventure's lack of conclusion was unfair on the part of the author, this type of "ending" is often the case for those conducting historical research. When St. Augustinians moved in 1763, the documentary record often stopped, leaving the scholar without further details regarding what happened to the individuals in question, as is the case for one of the families discussed here.

Despite these challenges, there is still a wealth of information for what life was like in the Spanish Americas. While St. Augustine has a rich 450-year-long history of European occupation, and an even longer history of Native American activities, this study concentrates on the eighteenth century. Consequently, the discussion below focuses on this time period. There are, however, numerous interesting and useful studies of both earlier and later periods (e.g., Cusick 1993; Lyon 1992; Manucy 1978; St. Augustine Historical Society 1955–2014; Waterbury 1983).

A Brief History of St. Augustine

St. Augustine's history began in 1565 when Pedro Menéndez de Aviles and Philip II, king of Spain, entered into a joint venture to establish a colony on the coast of Florida (Figure 5.1) and therefore defeat the encroaching

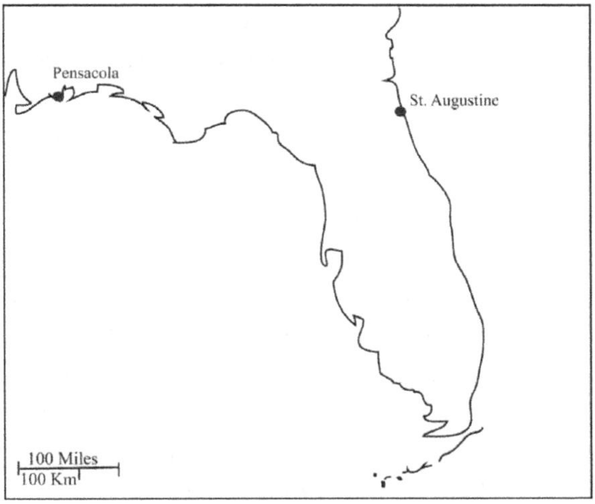

Figure 5.1. Map of Florida showing the location of St. Augustine. Map drawn by the author.

French colonists in the area (Deagan 1983:22). From the outset, the town was plagued by natural disasters and setbacks, including conflicts with local native tribes, hurricanes, floods, fires, disease, and attacks from the French, English, and pirates (Bushnell 1983:30–39, 53–55; Deagan 1983:22–23).

Despite these challenges, St. Augustine prospered, more than doubling in size during its first century (Deagan 1983:24). While the majority of the population were Spanish soldiers stationed at the local garrison, the town also included a diverse group of Europeans. By 1607 St. Augustine had 28 Portuguese residents, 20 French inhabitants, six Germans, and two Flemings from Belgium (Bushnell 1983:43). It is perhaps worth noting that, while the town's population was diverse, portions of all of these regions were under Spanish control at the time (Skowronek 1992), and these residents and their neighbors might have had different concepts of what it meant to be "Spanish." Someone born in Genoa during this period, for example, would have been a subject of the Spanish crown; however, that person might have self-identified as Genoese rather than as Spanish.

By the eighteenth century, the population had expanded to include Swiss individuals (not under the Spanish crown), several hundred Canary Islanders (Spanish subjects), Native Americans, and both free and enslaved Africans (Deagan 1983:31–34; Landers 1999). Mid-eighteenth-century baptismal and marriage records from the local parish indicate that people from other parts of Spanish America and Spain continued to immigrate to the town. Of these new arrivals, at least one individual was from Jerez de la Frontera, and another, Juan de Salas, discussed below, was from Santiago, Spain.

While this growing and diverse population was largely supported by the military presence in the town, scholars differ on how well provisioned St. Augustine was during its first two centuries. Traditional interpretations of the town describe it as a backwater, struggling frontier outpost that was barely able to obtain supplies because of the chronically late *situado*, or annual subsidy, from the Spanish crown (Deagan 1983:34–39). Although there were occasional challenges and problems with official supply channels, recent studies indicate that the town actually had a thriving, if not completely legitimate, economy (Lyon 1992:15–17). Evidence from sixteenth-century shipping and probate records indicates that the residents owned and had access to a wide array of goods from both Europe and the Americas (Lyon 1992), while archaeological studies indicate that St. Augustinians had plenty of food and, in some cases, were even able to maintain Iberian foodways (Reitz and Scarry 1985).

By the eighteenth century, residents of St. Augustine had successfully "diversified" their economy. Although they were technically breaking Spanish law, St. Augustinians frequently traded with nearby Havana as well as with their British-American neighbors to the north. From 1717 to 1739, the town's citrus industry supplied fruit to British-American ports, including Charleston, New York, and Philadelphia, in exchange for various goods (Harman 2004:22–23).

This diverse economy is reflected in both the archaeological and documentary records. Excavations throughout the city, including at the sites discussed here, have documented a high percentage of British and/or British-American ceramics in eighteenth-century contexts. In 1712 the sale of illicit British goods was so rampant that Governor Córcoles (held office 1706–1716) attempted to stop one of the more successful merchants by trying to arrest his second-in-command (TePaske 1964:83–84). The locals were so outraged that the governor would try to limit one of their sources of supplies that they pressured him into dropping the charges against the merchant (TePaske 1964:84). Ultimately, the town's residents turned to piracy; Governor Montiano (held office 1737–1747) himself hired privateers to capture ships carrying potentially useful supplies (Harman 2004:37–40; Waterbury 1983:79–81). By 1741, as many as 13 captured British ships rode at anchor in the town's harbor (Waterbury 1983:81).

In addition to lists of imported consumable goods (St. Augustine Historical Society Research Library 1759), city archives record that slaves were also moving between British and Spanish America, with at least one Philadelphia-born slave sold to a Juan de Salas in St. Augustine. Whether this is the same de Salas discussed below is unclear.

Although eighteenth-century St. Augustine was a relatively thriving urban frontier (Lyon 1992:16), in 1763 the Spanish crown ceded the town, along with other Spanish American territories, to the British as part of the Treaty of Paris following the Seven Years' War. As a result of this governmental shift, nearly all of the town's residents left, including the Ponce de León and de Salas families. Because they could not sell their houses before leaving, many St. Augustinians made an arrangement with Englishman and longtime resident of the town Jesse Fish to sell their property for them (Schafer 1983:96). While the majority of the residents moved to Havana, several also left for Campeche, Mexico. The actions of both groups indicate that they preferred to stay under Spanish rule and remain Catholic rather than convert to Protestantism and become British subjects. Given that St. Augustinians had a long history of interactions, both peaceful and violent, with British Americans, this mass exodus is telling of the residents' enduring ties to their cultural and religious heritage.

From a scholarly perspective, the migration is a significant point in the archaeological and archival record of St. Augustine. As families prepared to move and sell their property, they discarded a significant amount of material culture and thereby created numerous archaeological features that can be very tightly dated to within a few years of 1763. Additionally, the prospect of selling large amounts of property led individuals such as Fish, Thomas Jeffreys, and accountant Juan Joseph Elixio de la Puente to carefully document who owned which lots and what kinds of buildings were present. In 1764, for example, Elixio de la Puente drew an elaborate map on which he marked buildings, who owned them, what kinds of structures were on the properties, and lot sizes. Five years later, in 1769, Thomas Jeffreys drew a similar map that documented the locations of large structures, outbuildings, gardens, orchards, and streets. While Jeffreys did not record who lived at these houses, his map is useful in that it shows how the town changed between 1764 and 1769. These details, especially when combined with Fish's meticulous records about buying and selling the different lots in town, allow scholars today to connect eighteenth-century individuals with their property, how they might have improved or altered the lots, and how much each parcel of land was worth. This period of both household clearance and extensive documentation is a boon for those studying the end of the First Spanish Period (1565–1763), because it is relatively easy to identify property owners and connect them to the objects recovered archaeologically.

Although equally well documented, the British Period (1763–1784) was comparatively short-lived, because the Spanish regained control of the city in 1784. The Second Spanish Period (1784–1821) prompted another demographic shift when many St. Augustinians moved back and attempted to regain their property. By the 1820s the town again had a diverse population, including Spaniards, English and French residents, Germans, Italians, Greeks, Swiss, Americans, Minorcans, Canary Islanders, Scots, Irish, Native Americans, and Africans (Griffin 1983:130). In contrast to the mass exodus by Spaniards at the start of the British Period, close to three hundred English individuals chose to stay in the area and convert to Catholicism after the town returned to the Spanish crown. In 1821, however, the town's status as part of the Spanish Empire came to a permanent end when Florida became an American territory and, in 1845, a state.

Archaeology in St. Augustine

While St. Augustine's history made it a tourist destination as early as the 1840s, it was not until the 1950s that archaeologists began to take an interest in the city and its wealth of archaeological resources (Harrington et al. 1955; Buker 1983:171). By the 1970s Florida State University and the University of Florida had established a strong research presence in the town; students under the supervision of Charles Fairbanks, and later Kathleen Deagan, excavated a wide range of sites throughout the historic downtown area. In the late 1980s, the city instituted an ordinance that requires the city archaeologist to review any proposed construction projects, both private and public, prior to the start of construction, for their impact on the archaeological record (Appler 2012; *Code of the City of St. Augustine*, 2014). Thanks to the efforts of archaeologists from Florida State University and the University of Florida, the city archaeologist, and the St. Augustine Archaeological Association (SAAA), hundreds of sites have been excavated in the greater St. Augustine area, including the two sites discussed below (Figure 5.2).

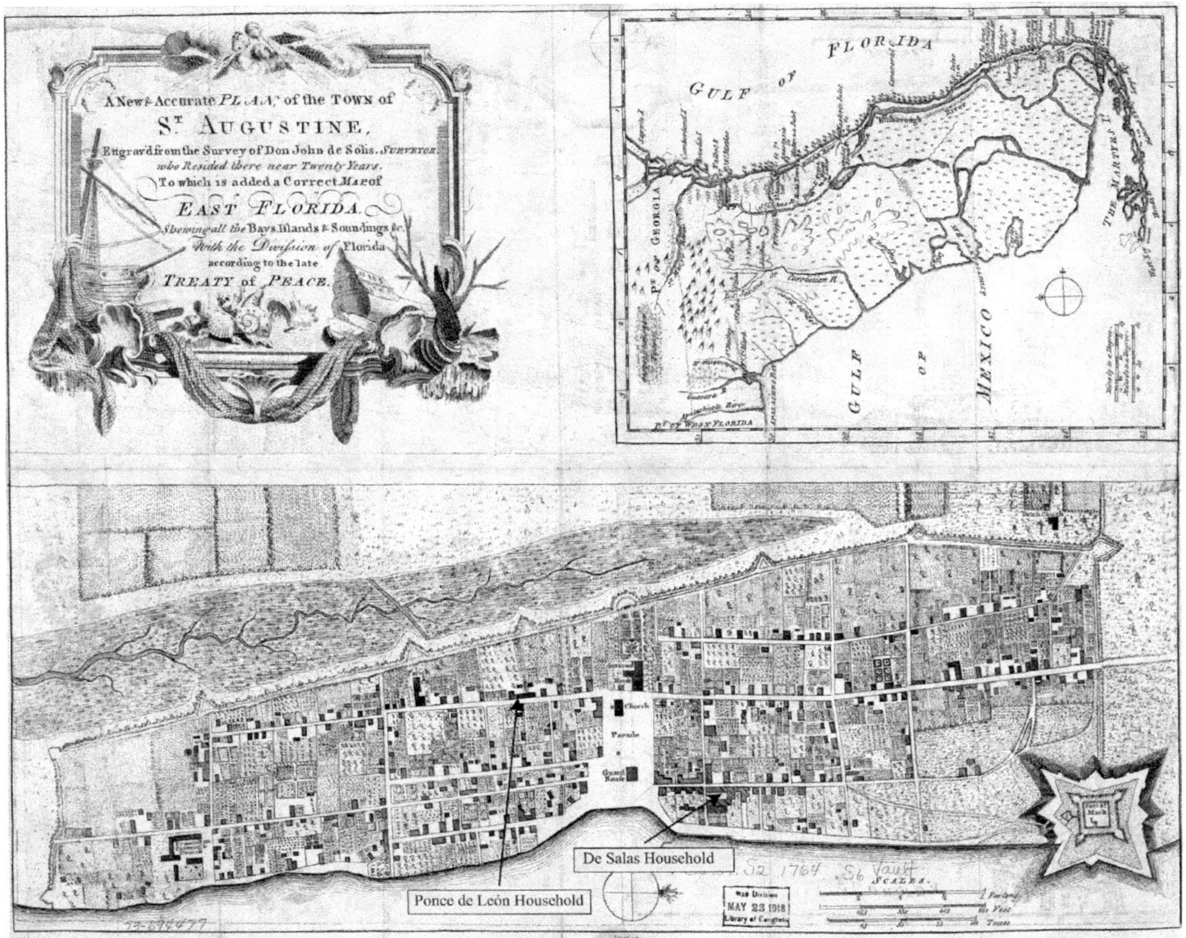

Figure 5.2. Copy of the 1764 Elixio de la Puente map. Image courtesy of the Library of Congress, Geography and Map Division.

Because of the diverse number of stakeholders, the artifacts from these excavations are held in several locations. Objects recovered from Florida State University and the University of Florida excavations currently reside at the Florida Museum of Natural History (FLMNH) in Gainesville, Florida, where scholars can access them for exhibits and research. Artifacts excavated according to the city's Archaeological Preservation Ordinance, however, belong to the landowner unless donated to the city. The donated items are under the care of the city archaeologist and are also used for research and exhibits.

The Ponce de León Household

The first Floridian household in this study belonged to Francisco Ponce de León and his family. The archaeological remains of their property, officially designated as SA 36-4, is known to archaeologists as both the Ponce de León Site and the Palm Row Site. Roughly two blocks southwest of the main plaza, the property is located on St. George Street and is bisected by pedestrian walkway Palm Row, which dead-ends into Cordova Street to the west (Figure 5.2). Today, Palm Row features six early twentieth-century houses on the south side and front "yards" with carports or garages on the north side of the walkway.

Although two sixteenth-century wells recovered on the property suggest it was occupied early in the town's history, the site's documentary records begin in the mid-eighteenth century when the town was transitioning from Spanish to British control (Florida Museum of Natural History n.d.). During this time, the area was known as Block Q (later as Block 36) and covered the property between today's King Street and Bridge Street. In the eighteenth century, the block was bounded to the west by a marsh that surrounded the western edge of the city and was divided among 17 different owners, according to the 1764 Elixio de la Puente map. Of these landholders, 11 had houses on their lots, making it relatively underdeveloped for a block within the city limits and close to the central plaza.

On his map Elixio de la Puente notes that lot number 221, today the Palm Row site, was owned by Don Francisco Ponce de León. Francisco's property measured 40 by 82 *varas* (approximately 111 by 227 ft, or 33.8 by 69 m) and had a *ripio*, or tabby, house on it. Although the location of the structures on the property differs between the 1764 Elixio de la Puente map and the 1769 Jeffreys map, both show that there were buildings on the lot in the mid-eighteenth century. The structures marked on these maps, as well as the presence of at least two wells and a significant amount of domestic trash indicates that the lot was residential and, according to the Jeffreys map, had space for a large garden or orchard behind the house. Given that there are no records of Francisco owning another parcel of land on which he could have lived, it seems likely that he and his family did live at the property at some point in the middle of the eighteenth century.

The Francisco Ponce de León Family

Just as the maps offer some information about the location of Ponce de León's property and house, local archives provide a wealth of information about Francisco's life and family. The Ponce de León family arrived in the Americas during the early days of Spanish activity in the region, and its many branches spread throughout Spanish America. Records indicate that there were multiple Francisco Ponce de Leóns at various locations and times in history, which can make documentary research complicated. The accounts referenced here discuss the same Francisco who owned the Palm Row site, given the age of the individual listed and the time of his residence in St. Augustine.

According to parish records, Francisco was baptized in the St. Augustine Cathedral on February 24, 1710, the oldest and only son of Don Francisco Ponce de León and Doña Cathalina Rodriguez. He had four younger sisters. As some of the many descendants of Juan Ponce de León, the conquistador who "discovered" Florida for the Spanish, Francisco and his sisters would have been part of a prominent, well-established *criollo* family and were most likely among the town's social elite.

While little is known about Francisco's early life, he appears to have joined the infantry when he was around 20 years old. One account lists an infantry private of the same name as early as 1730 (TePaske 1964:29). In this position he would have made approximately 132 pesos per year, giving him a higher salary than more than half of the town's population (Deagan 1983:44). Despite his relatively low rank, he was not only qualified to serve as the governor's secretary but also had at least 150 pesos to purchase the position (TePaske 1964:29). While a Francisco Ponce de León was offered this rank, he apparently did not want it and remained a soldier (TePaske 1964:29). If, as is reasonable to assume, this individual is the same Francisco Ponce de León who owned the property on Palm Row, then we can surmise that he was literate as well as comparatively wealthy and well regarded.

On December 25, 1743, Francisco married St. Augustinian *criolla* Doña Jacobina Pueyo. Between 1744 and 1763, the couple had nine children, all of whom were baptized in the local cathedral as their parents had been. Although researchers have suggested that the family had several slaves (Reitz and Cumbaa 1983:162), there are no clear records of these individuals in town archives.

According to his first child's baptismal record in October of 1744, Francisco was serving as an adjutant. In this capacity, his annual salary would have increased to 480 pesos, putting him within the top five percent of St. Augustinians in terms of annual income (Deagan 1983:44; Reitz and Cumbaa 1983:162). In 1763, when St. Augustine was ceded to the British, Francisco and his family joined the majority of their neighbors and left Florida (Florida Museum of Natural History n.d.).

After they left St. Augustine, their story seems to end. Contemporary records do not indicate where they went, nor does it appear that they returned in the Second Spanish Period to reclaim their property. Unlike his neighbors, Francisco did not grant his property to Jesse Fish or Juan Joseph Elixio de la Puente in hopes of having one of them sell it to the British newcomers for him. After the family left, their property passed through a variety of British and Spanish hands before being carved up into the six different lots that it is today.

Archaeology at the Ponce de León Household

In 1978 a Florida State University field school supervised by Charles Poe and Professor Kathleen Deagan excavated portions of the yards and carports of Palm Row 1 and 2 and recovered evidence of sixteenth-to-nineteenth-century occupation at the site. While the site provides data from the first century of St. Augustine's occupation as well as the British, Second Spanish, and American Territorial Periods, this study focuses on two eighteenth-century deposits: a trash pit known as Feature 8, and a well that was documented as Features 10, 11, 12, and 13 (Figure 5.3). The presence of creamware in these two deposits suggests that they were filled between 1762 and 1766, as this ware type has been found in other sealed contexts in Spanish St. Augustine that have been conclusively dated to this three-year window (Deagan 1975:21).

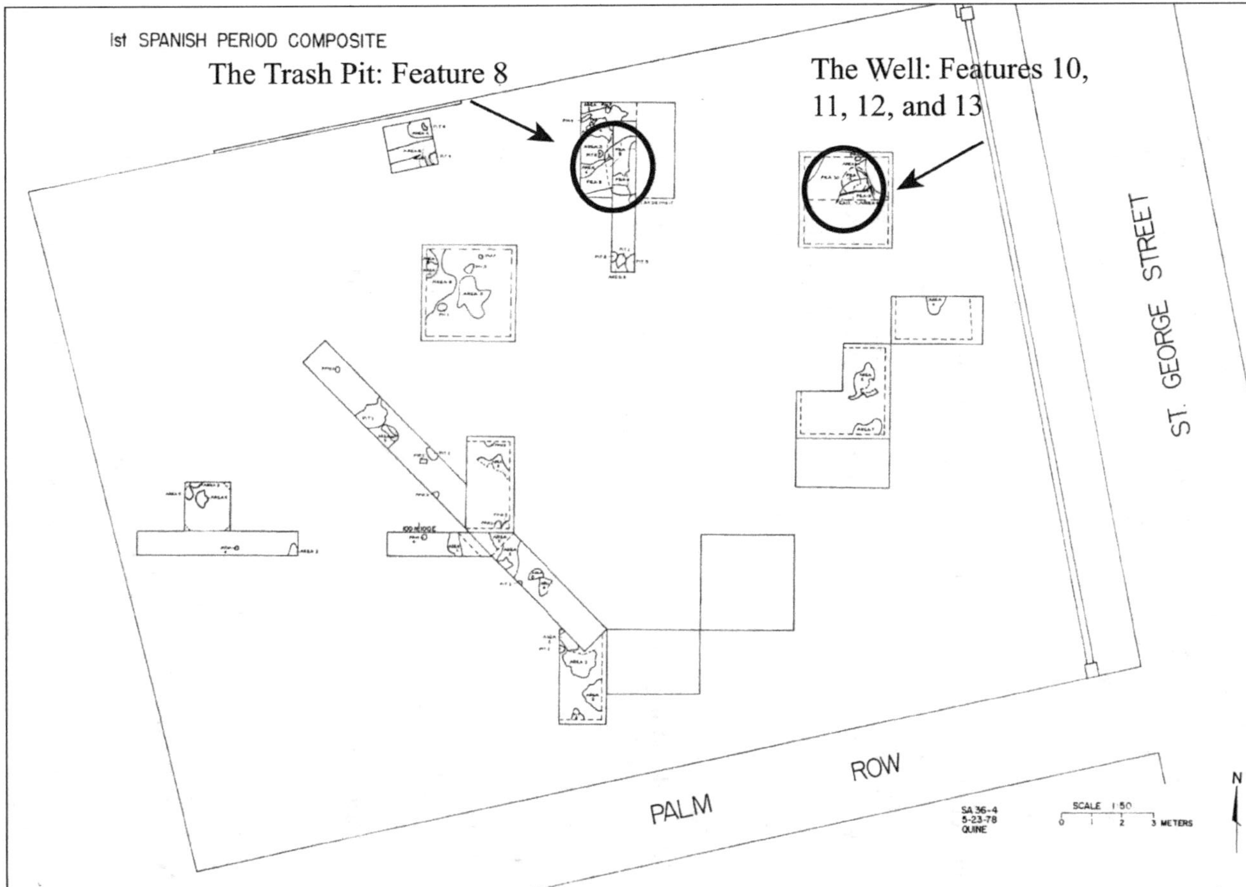

Figure 5.3. Map showing the locations of excavation units at the Ponce de León household. Image courtesy of the Florida Museum of Natural History–Historical Archaeology Collections.

This date range means the deposits were most likely the result of the household clearance activities that happened when the Ponce de León family prepared for their departure, along with most of their neighbors, in 1763. Because of the scale of the migration, it seems doubtful that the emigrants would have been able to take many household goods, such as dishes, with them, and therefore their preparations to move might have included discarding a significant proportion of their possessions. As a result, features like this trash pit and well provide an excellent idea of what one household owned and used at the end of the First Spanish Period.

The Trash Pit: Feature 8

The trash pit, Feature 8, was relatively large and filled with both faunal and cultural material. Although the present discussion focuses on the ceramics, previous studies of the faunal remains have yielded information about the foodways of the Ponce de León family and other *criollos* in St. Augustine (Reitz and Cumbaa 1983; Reitz and Scarry 1985). Crossmends among the ceramic fragments throughout the feature suggest that it was filled in a single episode (Ness 2015b). While the majority of artifacts recovered from it are of roughly the same date, it is important to remember that some of the earlier fragments might have been "echoes" of earlier activities at the site and were inadvertently included in the pit as the family members were disposing of their possessions.

The Well: Features 10, 11, 12, and 13

The other four deposits, Features 10, 11, 12, and 13, represent a well, its related construction activities, and fill layers. Based on the excavations, this well had both barrel lining and square, plank-lined walls and was filled in at least two different episodes (Ness 2015b).

Although there was relatively little cultural material in the well compared to the nearby trash pit, the types of ceramics from these deposits indicate that it was constructed in the mid-eighteenth century, while the similarities among the artifacts from the construction and fill layers of the well suggest that it was filled in shortly after its construction. It is unlikely that the family would have gone to the effort and expense of digging a well immediately prior to leaving St. Augustine in 1763, approximately when the trash pit was created. Consequently, it seems probable that the well was constructed in the mid-eighteenth century, shortly before the town was ceded to the British and the family members chose to emigrate. Once they made the decision to move, they would have most likely continued to use the well as they prepared to leave, filling it in only at the last possible minute with fragments from around the site as well as objects that did not go into the trash pit or other, unexcavated deposits.

Eighteenth-Century Features

Unlike La Calle Corredera, where the features offered an indication of activity areas at the site, the artifacts from the Ponce de León household were relatively homogeneous and included both utilitarian and tableware vessels. Indeed, nearly half of the ceramics from these eighteenth-century contexts were locally made, "aboriginal" ceramics that were used throughout St. Augustine as utilitarian vessels and cookware, with a total of 2,086 sherds falling into this category. Of these, 1,721 sherds were recovered from the well features, while 365 "aboriginal" sherds and all of the identified *loza común* vessels came from the trash pit. Unfortunately, it was not possible to determine an accurate minimum vessel count for these "aboriginal" ceramics, and they are therefore not included in the vessel counts for this study. The combination of utilitarian and luxury ceramics in all of these features supports the idea that the fill consisted of trash from the household that was most likely deposited during a site-cleaning episode either just prior to or just after the Ponce de León family left St. Augustine.

Regardless of the fact that the features do not clearly indicate activity areas at the site, they do provide information about the family's dining practices. While this will be discussed below and in the following chapter, it is important to note that the small size of the majority of sherds, particularly those from the well features, made it difficult to definitively identify some vessel forms. In many cases, it was possible to determine that a sherd belonged to an *escudilla* or *plato*, but it was difficult to distinguish whether or not it belonged in the *nuevo* or *viejo* category. Consequently, these more detailed categories have been used only when there was distinct evidence that a sherd belonged to a specific form rather than the more general categories (Figures 5.4 and 5.5; Ness 2015b).

The de Salas Household

The second St. Augustine household in this study belonged to Spanish merchant Juan de Salas. This property is part of a site known today as the

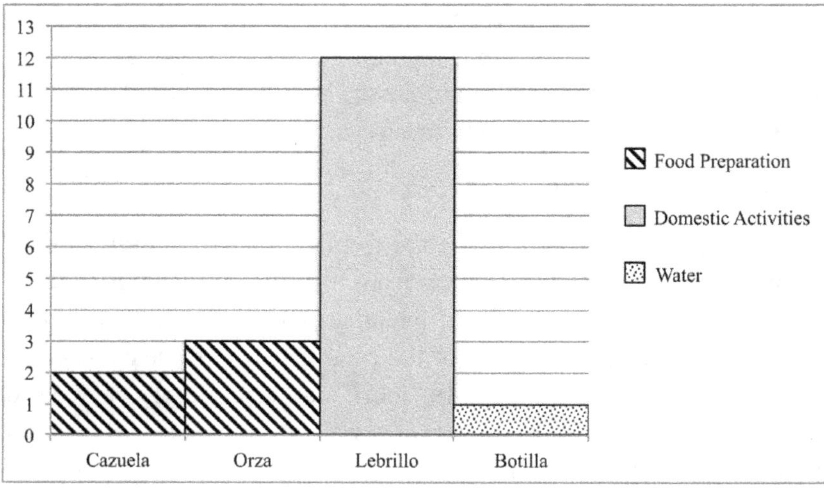

Figure 5.4. Minimum number of vessels (MNV) of *loza común* from the Ponce de León household.

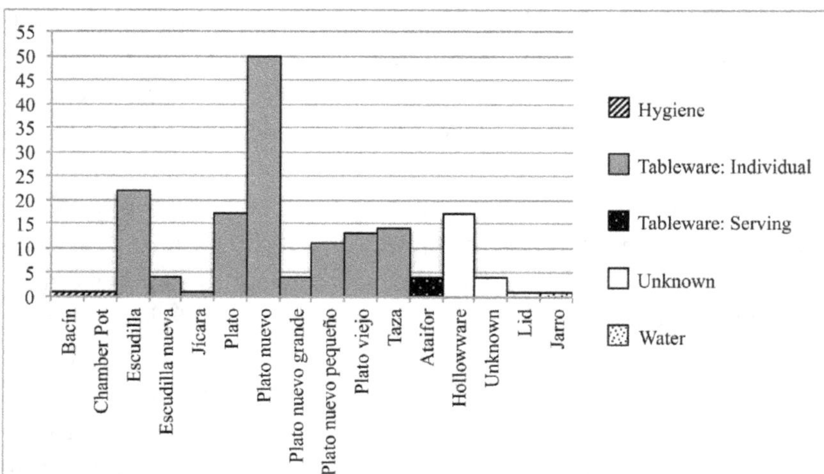

Figure 5.5. Minimum number of vessels (MNV) of *loza fina* from the Ponce de León household.

Monson Motor Lodge, officially designated as SA 3 and SA 4 or BDAC 99-0007 (Figure 5.2). Located between Charlotte Street and Avenida Menendez, this site is north of the main plaza and faces the Matanzas River to the east. The block is bounded by a pedestrian walkway known as Treasury Street to the south and the larger Hypolita Street to the north. Today, this property is the location of the Hilton St. Augustine Historic Bayfront Hotel, which opened in 2005.

As at the Ponce de León household, documentation of the ownership and residents of this site begins in the mid-eighteenth century. On Elixio de la Puente's map, the site was actually two blocks, L and M, and was divided in the middle by Baya Lane, a street that no longer exists but has been detected archaeologically (Halbirt 2003). Because Juan de Salas and his family lived on the southern end of the site and on the historic M block, this discussion focuses on the history and excavation of this portion of the site. Elixio de la Puente documents only four landowners on this small block, all of whom had stone houses.

In the eighteenth century, this section of Charlotte Street north of the plaza was the commercial center of town (C. Tingley, pers. comm., 2013). Soldiers traveling between the barracks at the south end of town and the fort at the north end would have traversed this street twice a day, making it a busy road and a prime location for merchants (C. Tingley, pers. comm., 2013). Additionally, the close proximity to the river and customs house made it easy for merchants to unload shipments and get merchandise to their stores (C. Tingley, pers. comm., 2013).

According to Elixio de la Puente, Don Juan de Salas owned lot 176 of block M. This lot was 15.5 by 47 *varas* (approximately 43 by 130.5 ft, or 13 by 39.7 m) and had at least one stone structure on it, presumably facing Charlotte Street. While some sources claim that de Salas shared ownership of his lot with his neighbor, Don Antonio Rodréguez de Arisan, and provide detailed descriptions of the structure (Arnade 1961:179; Parker 1999:94, 107), the building they describe appears to have been on the corner of Charlotte and Treasury Streets (Arnade 1961:179) and was therefore situated on lot 177, directly south of the property in question here. This information, as well as later property transactions for lot 176 that list only Juan de Salas, indicates that he was the sole owner of this property.

The Juan de Salas Family

As is the case with Francisco Ponce de León, there are at least three individuals named Juan de Salas who lived in St. Augustine during the First Spanish Period. While one was born in the 1600s, two lived in the eighteenth century, with one born in 1706 and the other around 1720 or 1730. These two eighteenth-century men are frequently conflated in documentary research. Fortunately, careful scrutiny of the age and familial relations of the individual mentioned in the texts helps to differentiate these two individuals.

Juan Antonio Artus de Salas, the Juan in question here, was apparently born in San Andres Figuerido, Santiago, Spain, as the legitimate son of Clemente Artus de Salas and Maria Suieris. Documents suggest that he was born around 1720 or 1730, although they do not indicate when or why he moved to Florida. His move may have been fiscally motivated; in St. Augustine he would have been considered a *peninsulare*, or native-born Spaniard, and therefore of the highest socioeconomic status, regardless of his socioeconomic situation in Santiago.

The first definitive mention of this same Juan de Salas in St. Augustine's archives occurred on Monday, December 30, 1754, when he married a local woman, Juana Sanchez Velaronse. Between 1756 and 1763, Juan and Juana had three sons: Joseph, Clemente (presumably named after Juan's father), and Julian. Local baptismal records also indicate that a Juan de Salas owned a slave named Miguel Francisco who was born in Philadelphia around 1739 and baptized in St. Augustine in 1749. While it is unclear which Juan de Salas this document refers to, it seems likely that Miguel Francisco belonged to the Juan who lived at lot 176. If so, this means that the Spaniard was in St. Augustine at least as early as 1749, if not earlier.

By 1763 Juan owned three lots in town: two on Charlotte Street (M-176 and K-156 on the Elixio de la Puente map) and another on Spanish Street (I-297). A fourth lot on Spanish Street (J-141) is also listed as belonging to Don Juan de Salas on the Elixio de la Puente map; however, this lot is not mentioned in other documents related to the *peninsulare* in question and might have belonged to the other Juan de Salas. In 1764, when the British assumed control, Juan, Juana, and their three young sons left for Havana along with the majority of their neighbors. He sold all three of his properties to Jesse Fish for 1,200 pesos, a significant sum when one considers that Don Rodriguez de Arisan, de Salas's neighbor and possible business partner, received only 200 pesos for his lot on Charlotte Street.

Archival records and historical studies from the First Spanish Period are unclear as to Juan's occupation and the cause of his financial success during that time. A 1766 list of St. Augustinians who moved to Havana includes a Don Juan de Salas listed under *vecino miliciano Españoles antiguos de Florida*, or old Spanish militiamen and residents of Florida (St. Augustine Historical Society Research Library 1766 AGI 87-5-3). While some have interpreted this source to suggest that he was a soldier (Arnade

1961:179), it is important to remember that all able-bodied men served in the militia, despite perhaps having another occupation (C. Tingley, pers. comm., 2013). Given that Don Rodriguez de Arisan was a merchant and ran a business from his house (Parker 1999:93–95), it seems reasonable to conclude that Juan de Salas may also have been a merchant, especially given the wide variety of goods recovered from his house and his ownership of such a well-located property.

Sometime between 1788 and 1793, Juan, his two younger sons, and one grandson returned to St. Augustine (Mills 1992:83, 102). Juan's wife apparently had died, and he was living with his son Clemente, also a widower, and three-month-old grandson on the south side of the plaza on what is today Aviles Street (Mills 1992:83). Julian, Juan's youngest son, seems to have been a master cooper who lived north of the plaza on Treasury Street, possibly on one of the lots Juan owned during the First Spanish Period (Archivo General de Indias 1787/1817; Mills 1992:102). Many families who returned to St. Augustine attempted to regain their old property, which might explain why Julian lived there.

While it is unclear if Juan successfully reclaimed his other lots, tax records document that by at least 1801 he owned a store and land in the Barrio de Iglesia, or Church Ward, north of the plaza (East Florida Papers 1801). Records indicate that he petitioned to have a billiards table in his store and that the business was shut down in 1799, apparently temporarily, for serving alcohol to Native Americans (East Florida Papers 1794, 1799). The fact that he owned and operated this store, however successfully, supports the conclusion that he was a merchant prior to his move to Havana. Juan lived the remainder of his life in St. Augustine, most likely working as a merchant until his death in 1807.

Archaeology at the de Salas Property

When Juan and his sons returned to St. Augustine, their former home, lot 176, was owned by English merchant John Leslie and was part of the Panton-Leslie Trading Company (Florida Master Site Files, n.d.). Despite being a British venture, this company was licensed by the Spanish crown after 1784 to remain in operation because it was considered an important factor in working with local Native American groups (Florida Master Site Files n.d.). The house, "which stood on the site from early times" and might have been the stone structure in which Juan once lived, burned in 1914 (Florida Master Site Files, n.d.).

Unlike the Ponce de Leóns' lot, which remained residential, Juan de Salas's former property ultimately became the site of the Monson Motor Lodge, a hotel where Dr. Martin Luther King and others were arrested during St. Augustine's civil rights protests in 1964 (Branch 1998:338). In the early 2000s, preparations were being made to demolish the motel and build the current Hilton St. Augustine Historic Bayfront Hotel, complete with an underground parking garage beneath most of historic blocks L and M.

In keeping with city ordinances, city archaeologist Carl Halbirt, his assistant, and a team of volunteers excavated sections of the area that would be impacted by this parking garage (Halbirt 2003). During this process they recorded a total of 200 features ranging from wall foundations and wells to trash pits and shell middens, although Halbirt speculates that this number is a small percentage of the features that existed on the site prior to the construction of the parking garage (Halbirt 2003).

One of the few eighteenth-century features uncovered during the construction process was a well, documented as Feature 199, that was on the southern edge of Juan de Salas's property (Figure 5.6). Unlike other circular wells nearby, this well was square and lined with a series of horizontal planks connected to corner posts (Halbirt 2003). Stratigraphic evidence suggests that the well had been cleaned at least once during its lifetime, while creamware in the fill layers means that it was filled between 1762 and the family's departure in 1764, before being sealed with a layer of clayey sand (Halbirt 2003; Ness 2015b).

The Well: Feature 199

Ceramics fragments from this well mend together to form vessels that were approximately 75 percent complete, making it possible to identify specific vessel forms (Figures 5.7 and 5.8; Ness 2015b). Because one would most likely not deposit such large pieces of trash down an active well, this find supports Halbirt's interpretation that the de Salas family intentionally used the well for trash disposal prior to moving to Cuba in 1764.

Like the Ponce de León household deposits, the de Salas family used "aboriginal" cookware, which was found in the lower and middle levels of the well. While utilitarian ceramics constituted nearly 50 percent of the ceramic assemblages from both La Calle Corredera and the Ponce de León household, only 12 percent of the ceramic fragments found at the de Salas household were "aboriginal" utilitarian ceramics, and the original

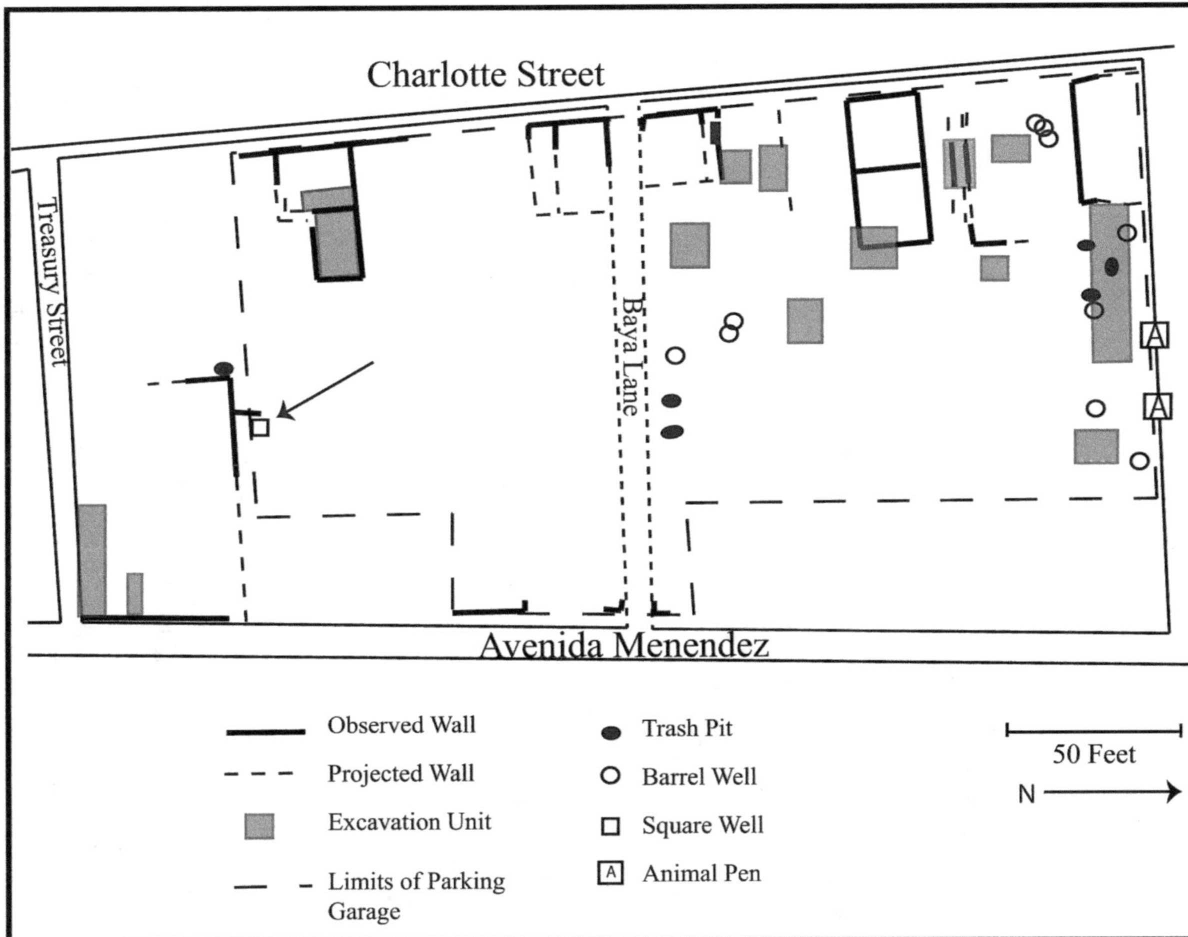

Figure 5.6. Map of the excavations at the de Salas household, with the well marked by an arrow. Image courtesy of the City of St. Augustine Archaeology Program.

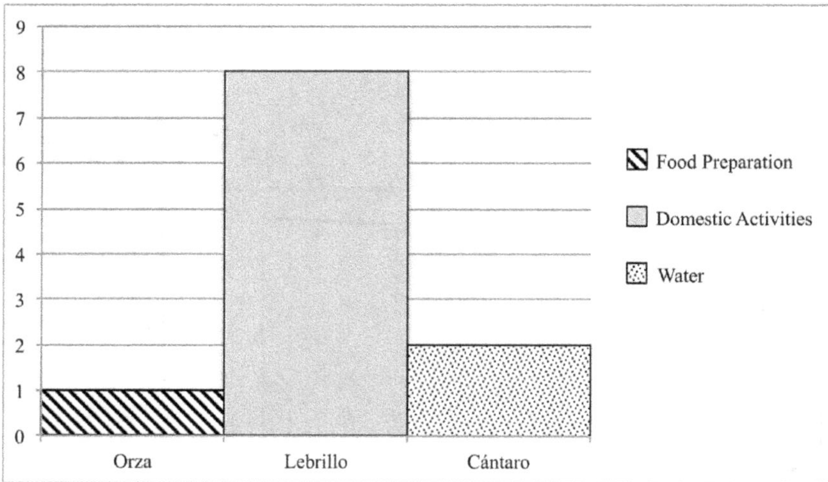

Figure 5.7. Minimum number of vessels (MNV) of *loza común* from the de Salas household.

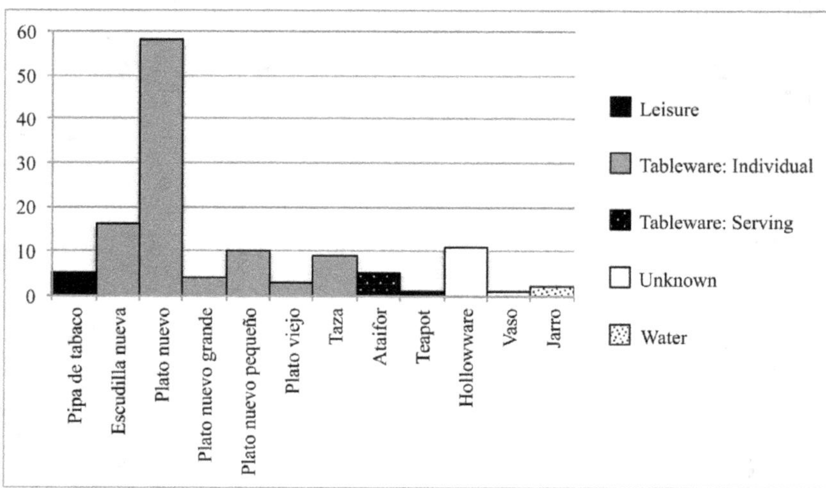

Figure 5.8. Minimum number of vessels (MNV) of *loza fina* from the de Salas household.

inventory documented only 119 sherds in this category. Because there were also relatively few *loza común* vessels at this site, the majority of the utilitarian ceramics might have been deposited elsewhere and were not systematically excavated.

Although the feature yielded a mixture of utilitarian and luxury ceramics, only nine percent of the ceramic assemblage were Spanish Mexican ceramic fragments, none of which mended with other fragments. Both the small size of these sherds and the lack of mends suggest that they were added to the well as part of cleaning up the site rather than through disposing of household goods.

Non-Spanish European and Asian ceramics, however, were abundant in all of the layers. The 58 *platos nuevos*, 16 *escudillas nuevas*, and nine *tazas* were made of creamware, delftware, white salt-glazed stoneware, slipware, and porcelain, including several matching plates. Many of the vessels were in multiple fragments and had clear mends, which indicate that a significant number of these vessels were whole or nearly whole when deposited.

The Ponce de León and de Salas Households

The two sites provide a great deal of data regarding *peninsulares* and *criollos* at the end of the First Spanish Period. According to the documentary record, the Ponce de León and de Salas families were financially well off. While we do not know de Salas's annual salary, the fact that he owned three lots in the town and sold them for an impressive 1,200 pesos in 1764 makes it seem likely that he, like Ponce de León, was among St. Augustine's financially elite.

The archaeological record supports this evidence that the two families were wealthy. Based on the ceramics recovered at the sites, both families were active participants in a global trade network that extended from Asia, Mexico, and British America to Spain, England, and the Netherlands. These artifacts and the general layout of the sites are representative of other elite domestic sites in town.

Where these two lots differ from other sites in town, however, is in the remarkable abundance and preservation of the objects recovered. As Halbirt (2003) mentioned, the well at the de Salas property is among the richest he encountered in his nearly two decades as the city archaeologist. Similarly, the trash pit at Palm Row was so noteworthy that Deagan (pers. comm., 2013) distinctly remembered it 35 years after it was excavated. Because of this preservation and high level of research potential at both sites, it is possible to learn about these families from what they left behind.

In all, these sites provide interesting insight into the consumption practices of the two families and, by extension, those of St. Augustine's elite. While studies (Carroll 1999; Henry 1991) advocate different scales for studying consumption patterns, ranging from the individual family level to the community, most agree that the term "consumption" refers to the acquisition of an object, the use of it, and the eventual discard of the same item.

Currently, it is difficult to determine exactly how the families obtained these objects, primarily because of gaps in the documentary information on these individual households and because merchant activity in St. Augustine during this period has yet to be studied in detail. As mentioned above, St. Augustinians had numerous options for trade, both legal and illegal, and might have obtained these goods through several different networks (Harman 2004). The situation is further complicated in that *rescate* and legal avenues might have contained the same items, making it virtually impossible without documentary evidence to accurately determine how an individual obtained an object. Consequently, while we know that St. Augustine residents illegally traded with non-Spanish sources, there is no conclusive evidence regarding how the Ponce de León and de Salas families obtained their dishes. Despite this uncertainty, we can glean a great deal about how they deposited them, as described above, and how they might have used them, as described below.

Setting the Table

The wealth of material from both sites makes it possible to establish a minimum vessel count of the tableware objects recovered and to gain an idea of what these families would have used at their tables. It is important to note, however, that the small size of many of the sherds makes it difficult to determine the exact ceramic type or shape of a vessel, especially with the sherds from the Ponce de Leóns' well.

In addition, the size of some sherds also made it difficult to determine the size of the vessel from which they came. While some scholars have suggested that early modern Spanish Americans used the term *platone* for large platters or chargers (Voss 2014), this term does not appear to have been commonly used in Spanish dictionaries or texts. Indeed, contemporary texts appear to be largely silent on the subject of size unless a specific vessel was designed to be large, as with *tinas*. Consequently, I have used *grande* and *pequeño* to note *platos* that were noticeably larger or smaller than the average-sized plate. In general, the *platos nuevos* had diameters ranging in size from 18 cm to 25 cm. Their larger counterparts, *platos nuevos grandes*, typically had diameters between 31 cm and 34 cm, and their smaller relations, *platos nuevos pequeños,* measured between 14 cm and 16 cm across.

The relative numbers of different vessel forms were similar at both sites, suggesting that the families performed similar activities and used similar items to express their identity. *Lebrillos*, for example, represent between 66 and 72 percent of the identified *loza común* vessels at the sites and, in both cases, were the most frequently found form within this category. Unfortunately, their multipurpose nature makes it difficult to determine precisely how these vessels might have been used at these households.

In addition to having similar *loza común* vessels, the two families owned and used similar percentages of tableware. At both households, between 30 and 40 percent of the *loza fina* were *platos nuevos*. They also had small numbers of *platos nuevos grandes* (~2 percent) and *platos nuevos pequeños* (~7 percent). Likewise, other tableware forms, such as *escudillas nuevas*, *tazas,* and *ataifores*, had similar percentages.

The ceramics indicate that both families also owned dishes with matching or semimatching patterns that they most likely used to create similar-looking sets. At the Ponce de León household, for example, the majority of high-quality tableware consisted of white salt-glazed stoneware that featured several similar, popular molded designs, like the "Dot, Diaper, and Basketweave" patterns. Several of the de Salas household's dishes featured similar decorations, although they appear to have had two sets of matching dishes, one creamware and one white salt-glazed stoneware.

Interestingly, these sets typically included only the *plato nuevo* and serving vessels such as *ataifores* and *platos nuevos grandes*. Other forms, such as *escudillas, tazas,* and *platos nuevos pequeños*, do not appear to have matched at either household and, within each household, seem to represent a variety of different ceramic types.

The situation is further complicated in that one vessel from the de Salas household appeared to be a deep, brimmed *plato*, or what modern Americans might call a soup bowl. If this form was popular, then it might have been how individuals ate soup or broth-based dishes at fashionable tables. Unfortunately, the rim fragments from this vessel are identical to those of plates found at the site and with similar patterns, making it difficult to distinguish with accuracy whether a rim or marly sherd belonged to a flat *plato nuevo* or this deeper variation.

Despite these similarities, there were several noticeable differences between the vessel forms found in the two ceramic assemblages. The Ponce de León household, for example, had nearly three times more *platos viejos* than the de Salas family, the majority of which were San Luis Polychrome (1650–1750; Florida Museum of Natural History 2012) and almost form their own set. These vessels could be interpreted in a variety of ways. The Ponce de Leóns' long history in the town would make it likely that Francisco or Jacobina inherited items like these dishes from their family members, unlike newcomer Juan de Salas. Alternatively, they could be older dishes that the family used before they purchased their more fashionable white salt-glazed stoneware set. One other possible interpretation of the *platos viejos* at this site could be that they were used by either small children or household staff, possibly slaves, who were not included in more formal dining activities and were therefore given less fashionable, potentially nonmatching dishes from which to eat.

The de Salas family also owned several nonmatching ceramics, although they were typically of the newer vessel forms (*platos nuevos* as opposed to *platos viejos*) and were made of European or Asian ceramics, suggesting that the children and household staff might have also enjoyed newer dishes even if they did not match the formal dining set.

It is perhaps worth noting here that these ceramic types might not have necessarily been more rare or expensive than other ceramics. Archaeological evidence from remote sites in today's Texas indicates that residents on the periphery of New Spain had access to a variety of Asian porcelains along with other types of ceramics (Longoria 2007). Archival studies, in turn, suggest that porcelain vessels, in some cases, were as expensive as Mexican-made vessels (Gasco 1992:85). Regardless of how much they cost the initial purchasers, the form is telling, as it suggests that all members of the de Salas household were eating non-broth-based foods prepared in new, possibly French, cooking styles.

While both households had small children, and possibly slaves, during the mid-eighteenth century, the differences in family composition also provide insight into how the ceramics might have been used. The de Salas household was half the size of that of the Ponce de Leóns, yet a similar percentage of its ceramics were *platos nuevos*, suggesting that the merchant and his family entertained on a more lavish scale than the Ponce de Leóns. This conclusion is supported by the fact that the de Salas family, unlike the majority of other families in Spanish St. Augustine, owned and presumably used, an Elers-type stoneware teapot, suggesting that the de

Salases could have also entertained both at full meals and at more informal, tea-party settings.

This vessel form and the implied activity associated with it might appear puzzling, especially given the traditional argument that the upper classes in Spanish St. Augustine, especially the *peninsulares* and *criollos* would attempt to seem as "Spanish" as possible (Deagan 1983), and tea is seen as a typically British beverage. I argue, however, that the presence of some traditionally non-Spanish forms, such as a teapot, at the de Salas household is actually in keeping with the attempt to appear Spanish. By the eighteenth century, chocolate, coffee, and tea were the most popular, nonalcoholic hot beverages in Spain (Pérez Samper 2009:25, 29; 2011:130). While chocolate and coffee were perhaps more widely consumed, there is some evidence that tea was also drunk throughout the Spanish Empire, even in relatively remote areas like Mission Santa Clara de Asís in California (Skowronek and Thompson 2006:16, 150). In Spain tea was common enough to be discussed in contemporary literature and was the subject of debates and medical experiments to determine its health benefits (Real Academia Española 1780; Lavedan 1796:134–200).

Regardless of one's perspective on the nutritional or medicinal aspects of tea, individuals consuming chocolate, coffee, or tea would have been expected to have the appropriate vessels from which to serve and drink these beverages. As a *peninsulare*, Juan de Salas would undoubtedly have been au fait with upper-class fashions in Spain and would have wanted to mimic them when he became wealthy in Spanish America.

Ceramic Types

This goal of staying abreast of current fashions in Spain is also reflected in the types of ceramics that the de Salas household enjoyed. Unlike the Ponce de Leóns, Juan de Salas and his family appear to have owned very few Spanish or Mexican ceramic vessels. Instead, they seem to have copied predominant trends in Spain by purchasing a significant number of British, British-American, Asian, and non-Spanish European ceramics. Because wear marks appear on both creamware and Asian porcelain sherds, it is possible to conclude that the de Salas family used these ceramics on a regular basis.

Given that the Ponce de Leóns, who also owned a significant number of non-Spanish European ceramics, were fairly representative of *criollos* in the town (Deagan 1983), one can hypothesize that other elite families in town were also adopting new ceramic types in an attempt to follow fashions in Spain. Consequently, this trend suggests that St. Augustinians were selective and did not always purchase items simply because they were available but rather sought out popular objects. While more research on the town's merchants and shops, including Juan de Salas's store, is needed to explore these ideas, the possibility that St. Augustinians were discerning consumers further undermines the notion that the town was a backwater outpost constantly on the brink of destitution.

The preponderance of English and non-Spanish European ceramics and the relatively low number of Spanish and Mexican ceramics in the de Salas household also appear to directly contradict the notion that ownership of non-Spanish ceramics was inversely related to income (Deagan 1983:238–239). While earlier Spanish colonial studies often equate ceramic types with socioeconomic status and cultural identity, one also must consider both what was fashionable and how the objects could have been used. In the case of the Ponce de León and de Salas families, they owned and used objects that were fashionable among elite circles in Spain, thereby demonstrating both their wealth and their sociocultural identity as members of the Spanish Atlantic community.

Conclusion

These two households provide a great deal of information about life in St. Augustine. The Ponce de Leóns' artifacts indicate that the family members were active participants in a global economy while also attempting to follow current Spanish fashions and trends. The de Salases' artifacts mirror these findings, although they seem to contradict the traditional hypothesis that wealthier Spanish Americans used Spanish and Spanish American ceramics to indicate their socioeconomic status.

From the luxury ceramics at these two sites, it is possible to conclude that the Ponce de León and de Salas families were aware of the trend in Spain and elsewhere to use matched sets and attempted to follow the fashion. They owned and used high-quality dishes made with newly developed techniques for hard, white-firing ceramics that were decorated with the latest designs. Those at the table were served from and ate from dishes that, if not exactly matching, were most likely aesthetically similar enough to be considered sets. If these sets included some *escudillas* or the deeper *platos nuevos*, then the diners most likely ate some soups and stews, although the predominance of flatter *platos nuevos* suggests their cuisine was largely non-broth based. When compared with sites from Spain, such as La Calle Corredera, these sites also reveal how those living in St. Augustine situated themselves within the broader Spanish Atlantic, which is discussed in the next chapter.

6

Expressing Spanish Atlantic Identity

Throughout *Don Quixote*, the hero encounters individuals who are preparing to travel to the Americas or who have relatives seeking their fortunes in the New World. As mentioned in chapter 5, for example, he meets and attempts to "free" a woman who is about to start a new life in the Americas with her husband. He also meets someone whose brother has moved to the Americas to seek his fortune. Last, the Knight of the Rueful Countenance has a friend, the curate, who inherited a substantial sum of money from a relative in the Indies (and who, incidentally, was robbed by a gang of thieves freed from prison by Don Quixote in an earlier adventure).

Tellingly, however, our chivalrous hero does not meet anyone who has actually returned from the other side of the Atlantic, only people who are traveling *to* the Americas or someone who has received money from the Indies. As a result, Don Quixote and Sancho have several bizarre ideas about the Indies and the passage there, suggesting that news and goods from the Americas were not always readily available in Spain.

Although the references to the Americas are subtle and easy for the casual reader to overlook, they are indicative of a strong cultural and academic concept: that colonialism is unidirectional. Studies of colonial situations often focus on the changes that occurred in the dominated territory and the adaptations and hybridity that occurred among those living there (Cusick 2000; Ewen 1991; Glasco 2005; Loren 2000, 2008). While colonized areas certainly experienced significant political, social, and economic changes, they were not the only places that were impacted by colonial activities (Cannadine 2001).

In both Spain and the Spanish Americas, there is evidence that fashions, social behavior, and foodways changed considerably during the eighteenth century. At least one Spanish writer observed that the majority of luxury goods in previous centuries were locally made, while eighteenth-century consumers preferred a wide variety of imported and luxury items (Cadalso 1796:106–107). Following trends seen in the Bourbon court, many individuals in both Spain and Spanish America welcomed new fashions and technology from Asia and elsewhere in Europe, especially England and France. Similarly, individuals on both sides of the Spanish Atlantic embraced new foods and goods from the Americas, which had a dramatic impact on Spanish foodways and material culture.

The cultural developments that result from the adoption of new ideas, technology, and products are phenomena well known among anthropologists and social scientists (Benedict 1934; Longhurst 2007; Steward 1976; Tax 1951). In culture contact and colonial situations, these changes are often closely related to notions of identity and self-authentication (Loren 2008; Voss 2005, 2008a, 2008b). This connection between cultural revolution and identity is particularly obvious in eighteenth-century Spain, as individuals grappled with defining what it meant to be "Spanish" while also attempting to negotiate social, political, and cultural changes (Fattacciu 2011:154).

The blend of American, European, Asian, and Spanish objects found at La Calle Corredera and the Ponce de León and de Salas households illustrates this struggle. The residents of all three sites were active participants in the global economy and sought to appear fashionable while preserving their Spanish heritage. In doing so, they expressed a Spanish Atlantic identity that combined new ideas and goods with traditional activities and settings.

Because of the increased globalization of the period, it is worth using a multidirectional perspective when examining eighteenth-century Spanish imperial activities and cultural developments. Ceramics often act as indicators for such cultural changes, as vessel forms and functions can shed light on activities such as entertaining and dining practices as well as modifications to both fashions and foodways.

Food Preparation and Foodways

Food was a powerful social marker in eighteenth-century Spain (Pérez Samper 2011:11). In previous centuries, displays of wealth often involved grand banquets with a vast array of courses. Eighteenth-century Spaniards, in contrast, sought to demonstrate social aspirations through more diverse and exotic goods (Cadalso 1796:106–107; Pérez Samper 2011:10). While the majority of Spaniards on both sides of the Atlantic might not have found their foodways changed dramatically between the seventeenth and eighteenth centuries, they gradually adopted American foods and French cuisine, both of which had significant impacts on dining practices (Pérez Samper 2011:12).

Of the foods that crossed the Atlantic to Spain, scholars have heralded the potato (*la patata*), tomato (*el tomate*), chili (*la guindilla*), paprika (*el pimentón*), and chocolate (*el chocolate*) as "fundamental new products" of the seventeenth and eighteenth centuries (Nestor Luján y Juan Perucho, as cited in Díaz 2005). While dictionaries do not reference chilies until the nineteenth century, seventeenth-century and eighteenth-century dictionaries document and describe the aforementioned foods as well as other American foods not on this list, including avocado (*el aguacate*), kidney bean (*la judía*), maize (*el maíz*), peanut (*el maní* or *el cacahuete*), pepper (*la pimienta*), and turkey (*el pavo*) (Covarrubias 1611; Real Academia Española 1726–1739; Rosal 1611).

Although Spaniards were importing American foods in sufficient quantities to draw the attention of individuals who might not have had direct contact with the Americas, such as those who assembled dictionaries, Spanish cooks appear to have incorporated new ingredients into mainstream cuisine at different rates. Seventeenth-century Spaniards embraced peppers, beans, turkey, and chocolate, although cookbook authors of the time referenced only turkey and peppers (Granado 1614; Martínez Montiño 1611; Pérez Samper 2011:130). In the eighteenth century, they began to include tomatoes in a variety of recipes, including sauces, fish dishes, and egg dishes (Altamiras 1758; Pérez Samper 2011:130, 133). While this imported ingredient was mentioned only five times in Altamiras's *Nuevo arte de la cocina*, it had become a staple in eighteenth-century Spanish cuisine and was the main ingredient of many meals (Pérez Samper 2011:133–138).

Spaniards also might have adopted the new foods in some areas earlier than in others. The 1611 dictionary by Francisco del Rosal, for example, defines the term *patata* as a type of parsnip "from Sevilla that lives in the Indies," suggesting that some American foods were initially more common in Andalucía, where they first entered Spain, than in other parts of the country. This definition also implies that, for some Spaniards, American foods were associated with Andalucían cuisine rather than American foodways, which might help account for the success of some foods and the failure of others.

Other popular American foods were most likely reserved for the upper classes initially by virtue of being expensive imported items (Pérez Samper 2011:130). Wealthy individuals throughout Europe, for example, regarded turkey highly (Kiple 2007:142), although the meat rarely appeared in early modern Spanish cookbooks. Similarly, the Spanish royal family enjoyed sweet potatoes in the eighteenth century (Pérez Samper 2011:132), but these do not appear in recipe books of the time.

Chocolate

Perhaps the most popular American food was chocolate, which Spaniards consumed exclusively as a beverage in the seventeenth and eighteenth centuries, despite its absence from cookbooks. This foodstuff was frequently mentioned in travel and personal accounts (d'Aulnoy 1692; Glas 1764; Labat 1730; Townsend 1791–1792). By the eighteenth century, chocolate consumption had increased in the Spanish Empire to the extent that it was accessible in smaller towns, rural areas, and relatively remote territories like California, leading to the creation of an entire industry of artisanal chocolate with specialists organized in guilds (Perissinotto 1998; Fattacciu 2009, 2011:155–171). Indeed, it had become common enough that, although not everyone could afford to consume chocolate on a regular basis, all levels of society could at least taste it (Fattacciu 2009:75).

Because Spaniards created specific equipment to make and consume this exotic beverage, chocolate is arguably one of the best examples of how changes in food impacted the ceramics of the time and can therefore be detected archaeologically. In order to prepare the chocolate drink, for instance, one would take a wafer or tablet composed of chocolate and other dry ingredients, such as sugar and cinnamon, and mix it with water in a tall vessel called a *chocolatera* using a beater called a *molinillo*. Once blended, the drink would then be poured into small *jícaras* that were most likely set in *macerinas*, deep saucers, for the consumer. These objects are an important example of cultural exchange in the Spanish Atlantic, as they not only facilitated the consumption of a new food in Spain but also indicated ways in which manufacturers adopted and internalized foreign products and terms.

While the traditional way of preparing chocolate in Mesoamerica involved pouring the liquid between two vessels, the Spanish found it more efficient to use a *molinillo* to blend the chocolate and water together and still achieve the same type of froth (Coe and Coe 2013:115). This tool has a solid "ball" with jagged edges attached to the end of a handle that is rubbed between the hands to blend the drink. While dictionaries do not specify the material used to make *molinillos*, surviving examples are typically made of wood. There were no *molinillos* at La Calle Corredera or the Ponce de León or de Salas households, possibly because of the perishable nature of the utensil.

Although the Spanish invented *molinillos*, *chocolateras* were a combination of Spanish and French developments during the seventeenth and eighteenth centuries (Coe and Coe 2013:115, 156–157). Initially, Spanish *chocolateras* were made from a range of materials, including silver, copper, brass, and pottery and were tall vessels with a wide belly and one handle (a *jarro*) (Real Academia Española 1726–1739). French manufacturers modified the handle to make it a straight rod at a right angle to the spout, with a hinged lid that had a hole in the middle to allow one to use the *molinillo* without spilling the drink (Coe and Coe 2013:156–157).

While earlier vessels similar to *chocolateras* have been depicted and recovered in Mesoamerica and the American Southwest (McNeil 2009; Washburn et al. 2011), archaeological remains of *chocolateras* appear to be rare on early modern Spanish sites, as none were recovered at La Calle Corredera or the Ponce de León household. At the de Salas's property, fragments of an English Nottingham-type stoneware jug suggest a form that resembles the narrow-necked, wide-bellied *chocolateras*, but the sherds do not mend into a whole vessel, so it is difficult to identify the precise form. Given that *chocolateras* were usually not made from English brown stoneware, like Nottingham-type stoneware, it seems likely that this vessel was a *jarro* rather than a chocolate pot.

While *molinillos* and *chocolateras* were European inventions to aid in the social ritual of preparing and serving hot chocolate, the means of consuming it strongly reflect the American heritage of this food. The term *jícara*, for example, comes from the Nahuatl or Aztec word *xicalli*, a vessel made from dried and hollowed calabash gourds used for a variety of purposes (Kiddle 1944; Norton 2006:672, 686). Early Spanish visitors to the Americas first encountered these vessels as chocolate cups and appropriated the term as a label for small, handleless cups designed for drinking chocolate and perhaps tea (Coe and Coe 2013:134–135; Kiddle 1944; Lavedan 1796:164;) and resembling some teacups of the time. Individuals in both Spain and Spanish America also drank chocolate from *coco chocolateros* (Machuca 2012), a small cup made from a coconut shell set in a silver frame that appears in museums more often than in the archaeological record. While *jícara*, *xicalli*, and *coco chocolatero* vessels were made from different materials and take different shapes, the fact that Spaniards were willing to adopt a new term and invent a new vessel form specifically for the consumption of the new beverage shows both the importance of the food and the impact of American products and ideas.

Jícaras appear in abundance at sites like La Calle Corredera, which yielded a minimum of 64 *tazas* (Figure 6.1). While the term *taza* includes a variety of cup forms, and it is difficult to distinguish exact forms from only small fragments, the lack of handles and the small diameter of the rims and bases suggest that the majority of these could have been *jícaras* and used for consuming either chocolate or tea. There were considerably fewer *tazas* at the St. Augustine sites, as the Ponce de León property produced only 14 *tazas* and the de Salases' well yielded only nine.

Interestingly, the majority of the de Salas family's cups were larger, made of slipware, and had evidence of handles, suggesting they were more like mugs than delicate *jícaras*. The lack of chocolate cups at the St. Augustine sites, however, does not necessarily indicate that the residents were abstaining from chocolate. They could have adopted the drink without the special equipage, which would mean that their *tazas* might also have been used for tea, wine, strong spirits, or other beverages. Indeed, the de Salas family owned a teapot, indicating that they were consuming popular beverages, despite the lack of teacups from this particular deposit.

Macerinas, the saucers that held the *jícaras*, were also an American product, although they were developed in the 1640s by the Spanish American Marqués de Mancera, Viceroy of Perú (Coe and Coe 2013:135). In an attempt to find a better way to drink chocolate in formal, court settings, he hired a Lima silversmith to create a small plate with a tall ring in the

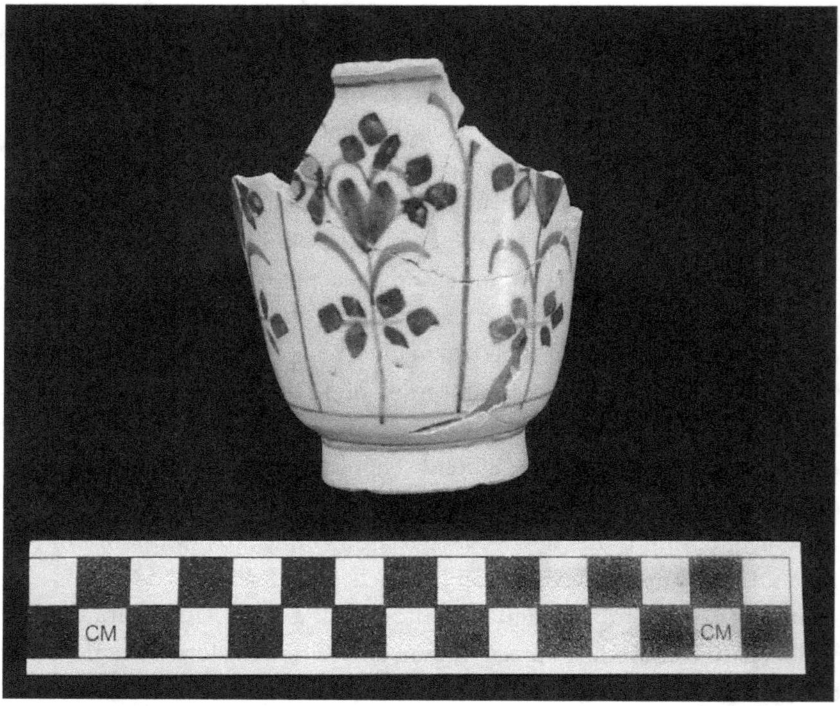

Figure 6.1. *Jícara* fragment from the *tinaja* (UE 82) at La Calle Corredera, deposited in the Museo Arqueológico Municipal de Jerez de la Frontera. Image courtesy of the Museo Arqueológico Municipal de Jerez de la Frontera and the Junta de Andalucía.

center that could act as a type of cup holder for the *jícara*. The vessel form was named after his title, Mancera, and was ultimately adopted in Europe and Mexico (Lister and Lister 1974), where it was made of both silver and ceramic.

While they are a well-known form and can be found in numerous museums throughout Spain and the Americas, *macerinas* appear to be fairly rare in the archaeological record and are not discussed in detail in archaeological literature. Of the sites discussed here, only one was recovered from La Calle Corredera, and none were found in the St. Augustine households. This form, however, is more difficult to identify than the other chocolate-related vessels, because its basic shape is similar to that of a plate, and its distinguishing feature is the deep cup holder in the center. If, therefore, one had only a rim sherd, either for the cup holder or the plate portion, one might misidentify the form as a cup, bowl, or plate.

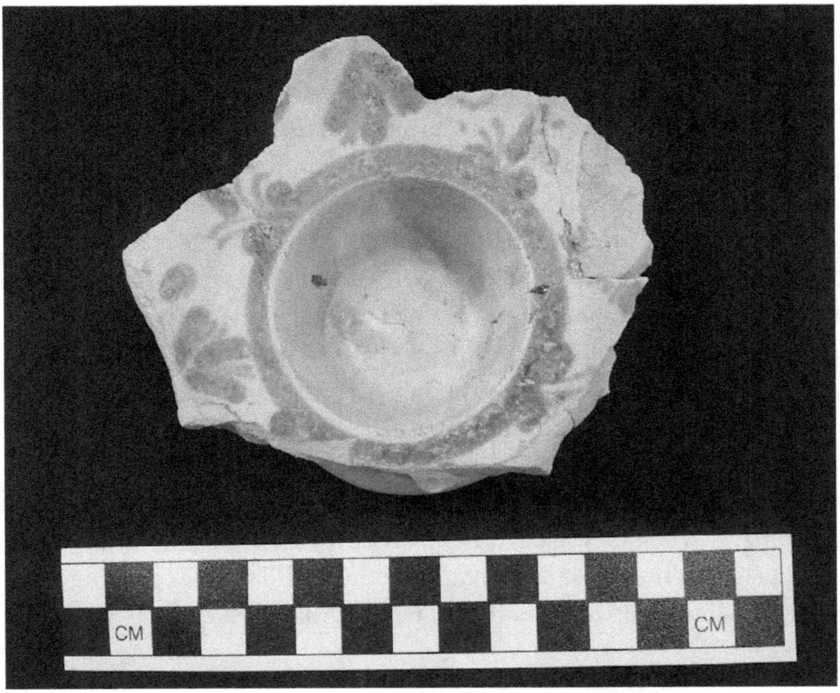

Figure 6.2. *Macerina* fragment from the trash pit (UE 54) at La Calle Corredera, deposited in the Museo Arqueológico Municipal de Jerez de la Frontera. Image courtesy of the Museo Arqueológico Municipal de Jerez de la Frontera and the Junta de Andalucía.

Consequently, one can definitively identify a *macerina* only if one has a large enough fragment or a sherd that includes portions of both the cup holder and the plate base, such as the one from La Calle Corredera (Figure 6.2).

Chocolate consumption provides an interesting insight to social behavior and expressions of identity. The development of special vessels for making and consuming chocolate, such as *molinillos*, *chocolateras*, *jícaras*, and *macerinas*, indicates that the drink was used as a symbol of one's status (Crown 2012:41). Unlike tomatoes and other American foods, this beverage required special preparation techniques, many of which were often associated with almost ritualistic activities and public display, as indicated by the high-quality materials used for *chocolateras*. Consuming chocolate was an equally significant social marker, because "anyone within the culture would recognize that someone, probably of high status, was drinking chocolate when they saw them lift their culturally appropriate form to their lips" (Crown 2012:41). Such social rituals were an important element of society, as they gave one a chance to show off affluence, genteel manners, and refinement (Rivas Pérez 2013:97–98).

The chocolate vessels at La Calle Corredera and the Ponce de León household as well as the de Salases' teapot, therefore, provide evidence that these families were refined and wealthy, or at least were attempting to appear so. They were sufficiently well informed about current fashions to know what beverages were consumed by elite individuals in Spain and other parts of Europe, and they knew there were certain rituals that accompanied chocolate and tea drinking. That they owned the specialized equipment to make and consume these beverages suggests that they were frequent chocolate and tea drinkers rather than more casual consumers who could not regularly afford the imported beverages or associated equipment. When they used these vessels, they demonstrated this knowledge as well as their economic status to others while also expressing their identity as members of fashionable Spanish Atlantic society.

Cazuelas and *Ollas*

While Spaniards developed specific vessels and utensils for chocolate, they incorporated other ingredients, such as tomatoes, into a variety of recipes that could have been prepared in, served in, or consumed from a variety of vessel forms. Consequently, many American ingredients appear to have had little impact on the ceramic material culture of the Spanish kitchen. As discussed in chapter 4, ceramic utilitarian vessels in Spain seem to have changed relatively little from the sixteenth through the eighteenth century. In the various features at La Calle Corredera, ceramic cookware is easily recognizable as either *cazuelas* or *ollas*.

In Spanish Florida, the majority of food preparation vessels were made of native ceramics that are typically unglazed and fired at low temperatures. The majority of the vessels were bowls, while some fragments from sites in St. Augustine had evidence of strap and loop handles, suggesting that the potters were also attempting to copy Spanish *ollas* (Otto and Lewis 1974:104–105). Unfortunately, the majority of "aboriginal" sherds at the Ponce de León and de Salas households were very small fragments or heavily eroded, making it difficult, if not impossible, to identify specific vessel forms. Based on evidence from other sites in town, however, it seems likely that these utilitarian vessels forms remained relatively consistent from the sixteenth through the eighteenth century, as was the case with utilitarian vessels in Spain.

Culturally, in both Spain and Spanish America, the *cazuela* and the *olla* were some of the most important items in the Spanish kitchen, as they were used to make the main courses. Linguistically, the words were important as well, because the terms *cazuela* and *olla* often referred to the vessel forms and/or the food that was prepared in them. This metonymical relationship is emphasized by the phrase *Después de Dios la olla* (After God, food), which the Real Academia Española's 1737 dictionary interprets to mean that, next to God, there is nothing greater than having food to eat (Real Academia Española 1726–1739).

While Altamira tended to use the term *cazuela* primarily in reference to the cooking vessel, other writers used the word to refer to its edible contents, and in Chile traditional Spanish stews are called *cazuelas* (Kiple 2007:185). In both Spain and the Americas, *cazuelas* were made from multiple recipes with a wide range of ingredients, including European, Mediterranean, and American foodstuffs. One recipe from a Jesuit cookbook, for example, is simply called *cazuela de tomates* and calls for fried onions, tomatoes, breadcrumbs, eggs, and spices like cumin and garlic (*La cocina de los Jesuitas* as cited in Pérez Samper 2011:136).

Although *cazuelas* were a common vessel form and meal, the stew called *olla* was traditionally the most fundamental daily fare for all levels of society and could be either sweet or savory (Pérez Samper 2011:61–69). Wealthier Spaniards might have had more or higher-quality meat in their savory *ollas*; all, however, ate some form of this dish, which included soup, meat, and vegetables, making it a complete meal (Pérez Samper 2011:61–62). While many recipes for *ollas* include primarily European or Mediterranean ingredients, such as garlic, onions, chickpeas, sausage, bacon, mutton, beef, chicken, and so on (Pérez Samper 2011:62–65), some recipes include tomatoes and peppers, indicating that New World ingredients were incorporated into this traditional dish by the eighteenth century (Altamira 1758). One common version was *olla podrida*, or "rotten pot," and was essentially a complex stew that included a variety of ingredients (Albala 2006:148). Interestingly, versions of this recipe have been found in cookbooks throughout Europe, suggesting that Spain was known for its stews (Albala 2006:148).

At La Calle Corredera, it is difficult to identify what exactly was being cooked in the 193 *cazuelas* and 94 *ollas* that were recovered from eighteenth-century features. Unfortunately, the excavation report and artifact inventory for this site do not reference any botanical studies, which means it is not possible to identify American plants such as peppers, tomatoes, or potatoes. The excavated faunal remains indicate that the family living at the site ate or had access to a variety of meats, including fish, shellfish, fowl, and large mammals such as cattle.

Such a varied diet has also been observed in St. Augustine, where residents relied on a mix of European plants and animals, such as figs, onions, garlic, cattle, pigs, and caprines, and wild American foodstuffs, including sweet potatoes, peppers, squash, maize, turkey, deer, crab, and fish (Reitz and Cumbaa 1983). Although St. Augustinians sent letters to Spain from the sixteenth through the eighteenth century to complain that they were frequently lacking in supplies and on the verge of starvation, the archaeological record as well as other documents suggests that these letters exaggerated the situation in hopes of obtaining better and more frequent supply shipments from Spain (Harman 2004:37–39; Lyon 1992; Reitz and Cumbaa 1983:152, 160). Indeed, faunal and botanical studies at sites throughout the town indicate that the residents had a rich and diverse diet that combined traditional Spanish cuisine with native cookery and, to a limited extent, African foodways (Reitz and Cumbaa 1983).

While the botanical remains taken from the excavations at the Ponce de Leon household have not been fully analyzed, faunal studies from eighteenth-century deposits at the site indicate that the Ponce de León family enjoyed a plethora of different meats (for a full species list, see Reitz and Cumbaa 1983:171–172). Approximately 80 percent of the biomass consumed by the family belonged to domestic animals, including pigs, cattle, and caprines, while wild terrestrial animals and birds constituted nearly 12 percent of the biomass recovered at the site (Reitz 1979:299). Aquatic reptiles, fish, and sharks made up slightly more than seven percent of the faunal assemblage from eighteenth-century contexts (Reitz 1979:299).

It is perhaps worth noting that, according to property descriptions and the Elixio de la Puente and Jeffreys maps, this property was bounded to the north by a swamp and was large enough to have an orchard and a garden. Consequently, the Ponce de León family may have made use of natural resources and grown at least some of their own food. While Francisco Ponce de León might have purchased wild meats or employed a hunting specialist, it is also possible that he hunted for sport (Reitz 1979). This idea is supported by the discovery of a fowling piece at the site and the fact that this family consumed a higher number of wild birds than did other contemporary households (Reitz and Cumbaa 1983:176–177).

Unlike the Ponce de León household, the de Salas household's foodways are relatively unknown, because the bone assemblage is small. The faunal remains that were identified, however, suggest that the *peninsulare* de Salas family ate a mix of fish and terrestrial animals, including shark, red drum, chicken, and cattle.

Just as Spanish Americans in St. Augustine adjusted their foodways by adopting local foods and using indigenous utilitarian vessels, some eighteenth-century Spaniards adjusted their foodways to incorporate new cultural influences from northern Europe. Recipes and accounts of food from the eighteenth century, for example, indicate that French cuisine became popular among elites in both Spain and the Spanish Americas after the Bourbons assumed the Spanish throne in 1700 (Kiple 2007:185; Pérez Samper 2011:64, 166, 338–342). According to English traveler Joseph Townsend (1791–1792:3:57), the archbishop of Granada had adopted "French cookery" by the 1780s, although the traveler does not elaborate on specific dishes or recipes that constitute this cooking style.

Scholars argue that while the Spanish royal court and those close to it attempted to imitate the pageantry and lifestyle of the court of Versailles, the average Spaniard initially noticed few changes in cooking and eating practices (Albala 2006:150; Díaz 2005:31; Pérez Samper 2013:338). While in Granada, for instance, Townsend (1791–1792:3:94) noted that the majority of people ate "after the Spanish fashion" which included the *olla* and other broth-based dishes, in contrast to the archbishop's French-style cuisine. Although it is difficult to know without botanical and faunal evidence what exactly was cooked and eaten at places like La Calle Corredera, the transition to *platos nuevos* at this site suggests that the residents actually ate fewer stews in the eighteenth century than in the sixteenth or seventeenth centuries and might have been mimicking elite Spaniards by attempting to prepare French recipes or by adopting French cooking techniques.

French culinary influence, while reflected in tableware, might also have been evident in nonceramic, utilitarian objects that are infrequently represented in the archaeological record. Just as *chocolateras* and *macerinas* were often made of silver, some kitchen equipment was made from metal or other perishable materials. Archival records from missions in California, for instance, indicate that individuals living there in the late eighteenth century used copper and brass pots, basins, dishes, kettles, and other vessels (Skowronek and Thompson 2006:66, 123, 148). Additionally, shipwrecks from the 1733 *flota* suggest that most of the utilitarian vessels aboard were copper (Skowronek 1984a:54, 59). Museum collections from the Museo de Costumbres y Artes Populares in Sevilla as well as the Museo Nacional de Cerámica y Artes Suntuarias "Gonzáles Martí" in Valencia also support the idea that ceramic vessels were often supplemented by a variety of nonceramic items, such as metal pans and tripods for down-hearth cooking, wicker baskets for storage, and wooden molds and utensils. Indeed, most of the kitchen objects found in reconstructed kitchens in Spain and Florida or in contemporary images of early modern kitchens do not appear to have been pottery.

Unfortunately, such nonceramic objects are not always present in the archaeological record, for a variety of reasons. Organic items, such as wicker baskets and wooden utensils, sometimes decay after deposition and, in a site like La Calle Corredera, are completely absent from the archaeological assemblage. Metal objects, in contrast, might survive as fragments but were more likely melted down and recycled. While there were some unrecognizable iron fragments at La Calle Corredera as well as brass buttons and tacks, there were very few metal objects at the St. Augustine sites, supporting the idea that the metals were reused rather than being discarded.

Alternatively, luxury objects like *chocolateras* that were made from precious metals might have been passed down in the family, and therefore would not appear in the archaeological record. Indeed, the minimal amount of metal of any kind at La Calle Corredera and the Ponce de León and de Salas households supports the hypothesis that metals were either reused or passed down through the generations.

Storage, Transport, and Nonedible Imports

In addition to foods, other nonedible goods were also being imported and exported throughout the Spanish Atlantic. Perhaps the most commonly discussed goods are the precious metals from America that transformed the Spanish economy (Cadenhead 1960; Haring 1915; Jones 2002; Thomas 2003), and the Spanish ceramics that were sent to the Americas (Deagan

1987; Fairbanks 1973; Florida Museum of Natural History 2012; Goggin 1968; McEwan 1989). Museums such as the Museo de Américas in Madrid and the Archivo General de Indias in Sevilla, however, have numerous items from the Americas that were sent to Spain as "curiosities," such as native clothing and art, feathers, and ceramics. In addition, individuals in both Spain and Spanish America clearly imported items and ideas from other parts of Europe, resulting in even more cultural blending.

Ceramics

Many imported items would not appear in the archaeological record, either because they were passed down in the family—such as objects made from precious metals—or too fragile, for example, feathers. Ceramics, however, do survive and provide excellent insights into imported items in both Spain and Europe. Additionally, these goods were often available to the average person and suggest that foreign objects and ideas were widely adopted in the Spanish Empire regardless of social status.

Although much has been published on European ceramics in the Americas, especially in St. Augustine (see, for example, Deagan 1987; Fairbanks 1973; Florida Museum of Natural History 2012; Goggin 1968; Skowronek 1984b, 1992), relatively little has been written about the ceramics that were sent from the Americas back to Spain (Logan 1977; Skowronek 1984a).

The two most obvious examples of American ceramics sent to Spain are types known in the Americas as Mexican Red Painted ware and Guadalajara Polychrome. The former ceramic is a relatively coarse, sand-tempered reddish earthenware with a smooth red exterior that is either burnished or painted (Florida Museum of Natural History 2012). Guadalajara Polychrome, also known as Aztec IV, is a fine grey earthenware that is burnished or polished and hand painted with black, red, brown, yellow-orange, and blue decorations (Florida Museum of Natural History 2012).

While these types come in a variety of forms, many of them are open or hollowware, and those that are closed forms are often too small to be storage containers. The collections from the Museo de Américas in Madrid, for example, includes several elaborate Guadalajara Polychrome/Aztec IV in varying shapes, ranging from *escudillas nuevas* and *tazas* to jars and open bowls with fluted sides, many of which feature small, ornamental handles. The shapes, combined with the elaborate decorations and the

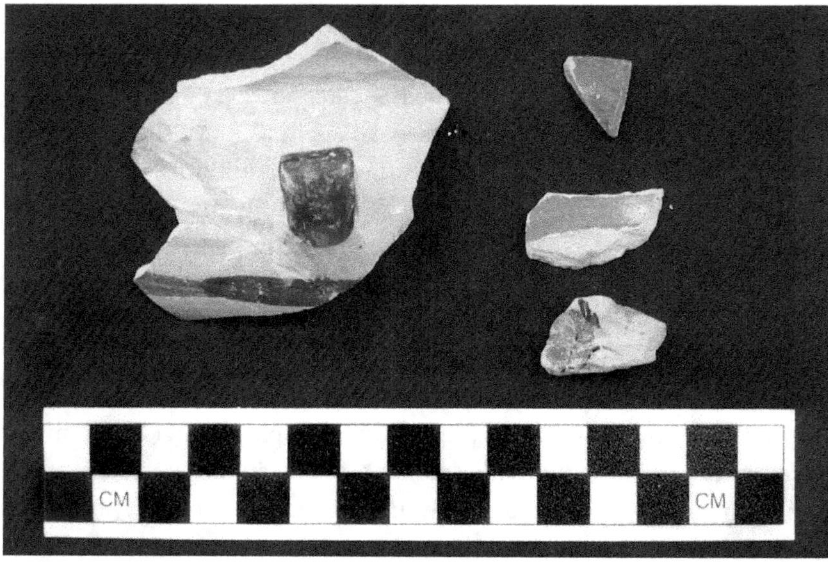

Figure 6.3. Guadalajara Polychrome/Aztec IV fragments from the *tinaja* (UE 82) at La Calle Corredera, deposited in the Museo Arqueológico Municipal de Jerez de la Frontera. Image courtesy of the Museo Arqueológico Municipal de Jerez de la Frontera and the Junta de Andalucía.

lack of a hard, durable glaze, suggest that these vessels were not subject to everyday use but rather might have been viewed as curiosities or decorative luxury items.

Examples of either Mexican Red Painted ware or Guadalajara Polychrome/Aztec IV were recovered at all three households. Among these sites, Mexican Red Painted ceramics were more common in St. Augustine, with 33 fragments found at the two households and none identified at La Calle Corredera. Guadalajara Polychrome/Aztec IV, in contrast, was apparently more popular in Spain than in Florida, as 18 sherds from at least three vessels were found at La Calle Corredera, while only one sherd was found in the St. Augustine households.

The presence of Mexican ceramic types at these sites indicates that the families were able to afford and had access to imported luxury items that might not have been overly functional. At La Calle Corredera, for example, two of the three Guadalajara Polychrome/Aztec IV vessels found appear to have been small *ollas* with wide bellies, narrow mouths, and flared lips (Figure 6.3). They were clearly not used for food preparation, however, as they had an unglazed interior and an elaborately decorated exterior with no evidence of burning.

Other sites in the Americas, including some in California (Skowornek et al. 2014:277–279) suggest that other Mexican ceramic types were also widely exported, possibly on a regular basis. Approximately 270 km south of St. Augustine, at the Higgs Point site, also known as 8IR00024, archaeologists have identified the remains of at least one Europe-bound ship that carried Mexican-made ceramics. This site dates between 1700 and 1725 and includes both a wreck and a small camp that was occupied by the survivors or salvagers of the wreck. Despite being looted prior to its excavation in the 1940s, the site yielded numerous objects that might have been on this ship, including some "Mexican majolica and nonmajolica ceramics" (Florida Master Site File 2003; Rouse 1951:214).

While ceramics from the Higgs Point site could have been from the salvager's camp rather than the shipwreck, other wrecks with similar cargo suggest that American ceramics were frequently bound for Spain or other parts of the Atlantic. Artifacts from 1733 *flota* shipwrecks, for example, indicate that Spaniards brought, or intended to bring, several kinds of Mexican majolica to Cádiz, including Puebla Blue on White and San Augustin Blue on White (Skowronek 1984a:43, 177). Although such vessels were present in comparatively small quantities on some of the ships (Skowronek 1984a:177), Mexican-made Puebla Blue on White "was the major majolica ware recovered from the wreck of the *San José*" ship in the 1733 *flota* (Logan 1977:29). While these vessels may have been intended for shipboard use rather than as cargo, the fact that they were on the ship indicates that at least some Mexican majolica was being transported to Spain and might have been familiar to Spaniards.

The idea that some Mexican majolicas were widely exported is also supported by historical archaeologist Charles Fairbanks's 1975 study of Spanish ceramics at Fortress Louisbourg in Cape Breton, Nova Scotia, specifically his identification of Puebla Blue on White vessels at this site. While he noted that the decorative motif does not match examples of this type found in the Caribbean, the "paste, form, and other features seem to fall within [the type]" (Fairbanks 1975:44). The two *ataifores* he illustrates in his article closely match at least three *escudillas nuevas*, two *ataifores* (Figure 6.4), one *plato nuevo*, one *jícara*, and the *macerina* from La Calle Corredera.

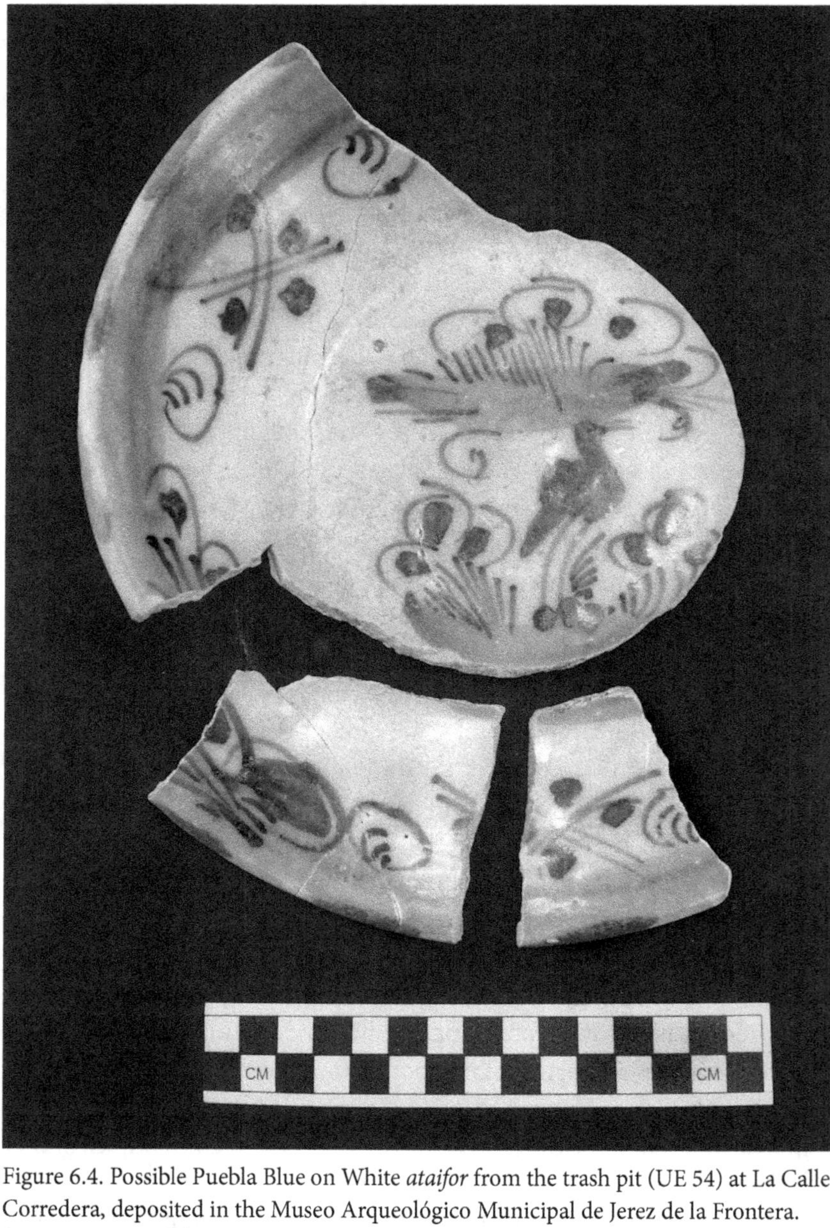

Figure 6.4. Possible Puebla Blue on White *ataifor* from the trash pit (UE 54) at La Calle Corredera, deposited in the Museo Arqueológico Municipal de Jerez de la Frontera. Image courtesy of the Museo Arqueológico Municipal de Jerez de la Frontera and the Junta de Andalucía.

Additionally, similar vessels were found during a salvage excavation in the historic town center of Rota, Spain (Gutiérrez López 2001:125). While elemental analyses of the ceramics from La Calle Corredera, Rota, and Fortress Louisbourg are necessary to conclusively determine if these vessels were made in Mexico, the ones recovered in Spain do not match other eighteenth-century ceramics from Triana, Sevilla, in that the decorations are much looser and less detailed, and the glaze is somewhat thinner than on locally made vessels.

If Fairbanks's assessment was correct and the vessels are from Mexico, then the presence of so many at a place like Jerez, which was not as directly connected with American trade as Cádiz, suggests that a significant quantity of Mexican majolicas were being exported. As a result, these vessels would have been available to a wide range of Spaniards and even non-Spanish individuals as far north as Nova Scotia. If, however, these vessels are of Spanish origin, then it could merit a reanalysis of the Puebla Blue on White ceramic type as it occurs in the American assemblage, as some ceramics previously identified as Puebla Blue on White might be Spanish rather than Mexican.

European Influences: Fashions, Aesthetics, and Tableware

Just as the Spaniards were importing Mexican ceramics like Guadalajara Polychrome/Aztec IV and possibly Puebla Blue on White, they were also adopting cultural influences from other parts of Europe. While Spain had long engaged in trade with England, France, Italy, and the Netherlands (Stevens 1702), the advent of the Bourbon dynasty in 1700 ushered in a new era of non-Spanish European influence in the Spanish Empire.

One traveler in Andalucía noted that by the early 1700s, the region had over 20,000 Frenchmen living there as laborers and servants (Labat 1730:286). While it is unclear how many of these individuals, if any, contributed to the artistic and cultural atmosphere of the time, other foreign influences from England, Italy, and France heavily impacted art, music, clothing, jewelry, architecture, and philosophy in Spain (Muller 1972:148). Eighteenth-century Spaniards began to read works by Enlightenment thinkers from France and England, and some families sent their children abroad either for education or, in the case of a very small number, for a grand tour (Herr 1958:73–74). Household inventories during the Second Spanish Period in St. Augustine indicate that residents there also had access to current Enlightenment treatises on philosophy and science (Cusick 1993, pers. comm., 2013).

While some Spaniards went to other parts of Europe for education, others went to observe the prevailing fashions and returned to Spain attempting to imitate what they had seen and heard abroad (Herr 1958:74). These individuals were often seen as somewhat effeminate dandies known as *petimetres* or *petimetras* ("petits-maîtres") and were frequently ridiculed by their countrymen, who saw their French affectations, both linguistically and sartorially, as degrading to Spanish culture (Herr 1958:74; Noyes 1998:199).

As mentioned above, this adoption of French fashions and language in Spain sparked an identity crisis that led to a rise in local traditions, especially in dress. Individuals such as Maria del Pilar Teresa Cayetana de Silva, Duchess of Alba, for example, often eschewed the French fashions predominant at court and chose to dress in the *majo* fashion, imitating the traditional, working-class attire of individuals in Madrid and the larger Andalucían cities (Noyes 1998:199, 210).

Although some members of the nobility and royalty rebelled against the crown's attempts to "bring Spain back to European standards" by promoting national fashions and dressing in clothes that were considered below their station (Noyes 1998:198, 208–210), French influences continued to pervade Spanish culture. Travelers in Spain often noted this rise of French fashions when they commented on individuals who prided themselves on their "French manners," clothing, or activities (Labat 1730:9; Townsend 1791–1792, vols. 2 and 3). Indeed, Townsend (1791–1792:2:141) observed that at the Duchess de Vauguion's house in Madrid, the entertainments were primarily French, including cards, chess, and *trictrac*, or backgammon. By the 1800s such customs were so highly regarded that at least one etiquette book, *El hombre fino*, was translated from French into Spanish; it includes very detailed rules regarding behavior in a variety of settings, including an amusing anecdote regarding the use of the relatively new fork (Rementeria y Fica 1837).

Just as elite society in Spain sought to mimic French fashions, Spanish artisans and potters were also influenced by their northern neighbors in several ways. After kaolin and petuntse were discovered in Saxony, and a royal ceramic factory was established in Meissen in 1710, potters throughout Europe began to create both hard- and soft-paste porcelain (Ray 2000:171). By the 1760s English creamware had also gained popularity in

Europe, and Spanish potters worked to find ways to create similar kinds of ceramics using local ingredients (Ray 2000:171). Hard, white-firing ceramics, such as porcelain and creamware, became highly sought-after among European consumers, with porcelain seen as "the necessary adjunct to gracious living" (Ray 2000:171). Like others in Europe, many Spanish ceramic factories attempted to produce their own porcelain and creamware, or *tierra de pipa*, and thereby supply local consumers with high-quality, popular ceramics. Two factories in particular led the field in these efforts: La Real Fábrica de Alcora and Buen Retiro.

In 1727 Don Buenaventura Pedro de Alcántara Jiménez de Urrea y Abarca de Bolea, the ninth Conde de Aranda, established La Real Fábrica de Alcora ceramic factory near Valencia in Alcora, Spain. His goal was to emulate the output of popular ceramic factories in Europe and thereby weaken the market for imported pottery in Spain (González Martí 1936:150; Ray 2000:171, 213). To do so, he hired French potters from Moustiers and Marseilles to train local workers, and his successor, the tenth Conde de Aranda, encouraged potters to experiment with porcelain and creamware manufacturing (González Martí 1936:155; Ray 2000:171, 220). While the factory created blue-on-white pieces inspired by Chinese porcelain for the first three years, it added polychrome decorations by 1730. Alcora ceramics were imitated by potters throughout Spain (Ray 2000:171, 215), especially the early, lacy Bérain-style motifs as shown in chapter 4 (Fig 4.9). For over 100 years, La Real Fábrica de Alcora set the standard for high-quality ceramics in Spain, and it has been widely studied by ceramicists, archaeologists, historians, and art historians as an excellent example of both an eighteenth-century Spanish entrepreneurial enterprise and a leader in eighteenth-century Spanish ceramic production (Ray 2000:213–239; Todolí 2008).

While the Alcora factory produced plaques and figurines as well as functional pieces such as *platos nuevos*, *escudillas nuevas*, *bandejas*, *macerinas*, and *jícaras*, the Buen Retiro ceramic factory at the Retiro Palace in Madrid created primarily ornamental pieces and even produced an entire ceramic room for the royal palace at Aranjuez (Ray 2000:171). Founded in 1760, this factory was also influenced by French fashions and ultimately developed a formula for a type of "hard-paste porcelain," although both the Alcora and Buen Retiro factories found it impossible to match the quality of imported porcelain or English creamware with local clays (Museo Arqueologica Nacional 2014; Ray 2000:171; Townsend 1792:3:231).

Despite its success among the elite and its reputation for producing high-quality decorative objects, the Buen Retiro factory apparently did not have as lasting an impact, nor was it as widely imitated, as the factory in Alcora. The discrepancy in the popularity of the two manufacturing centers is evident at La Calle Corredera, as there are numerous vessels from Alcora or factories that copied Alcora, and no obvious fragments from Buen Retiro.

Although it is somewhat more difficult to identify distinctly French influences in St. Augustine, the impact of European fashions is obvious among the ceramic assemblage. At both the Ponce de León and de Salas households, over 50 percent of the tableware recovered was of non-Spanish European origin. Delftware, creamware, and white salt-glazed stoneware were among the most popular ceramic types and represented a very wide range of vessel forms, ranging from *platos nuevos* and *escudillas nuevas* to *tazas* and *ataifores*.

It is important to remember that the features at La Calle Corredera were most likely filled in around 1740, approximately the same time that the white salt-glazed stoneware patterns popular in St. Augustine were being produced (Florida Museum of Natural History 2012) and nearly 20 years before creamware was invented. Because La Calle Corredera predates the St. Augustine sites by nearly two decades, the ware types are not completely comparable. The small Whieldon ware sherd (Figure 6.5) found at the Spanish site indicates that *jerezanos* had access to fashionable English ceramics. Likewise, other Spanish sites indicate that creamware was imported to Spain (Isbert and Andreu 2006), and by 1783 the English Leeds creamware pattern book was translated into Spanish, among other languages, and marketed to a non-English audience (Ray 2000:171).

Consequently, it seems likely that the family at La Calle Corredera might have been replacing their household dishes with something more fashionable, such as white salt-glazed stoneware, when they filled in the trash pit (UE 54) and the *tinaja* (UE 82). Likewise, the increased presence of British ceramics in St. Augustine suggests that individuals there might have also been targets of the remarkably effective marketing strategies of British potters like Josiah Wedgwood who were able to partially shape fashions based on their products and advertising campaigns.

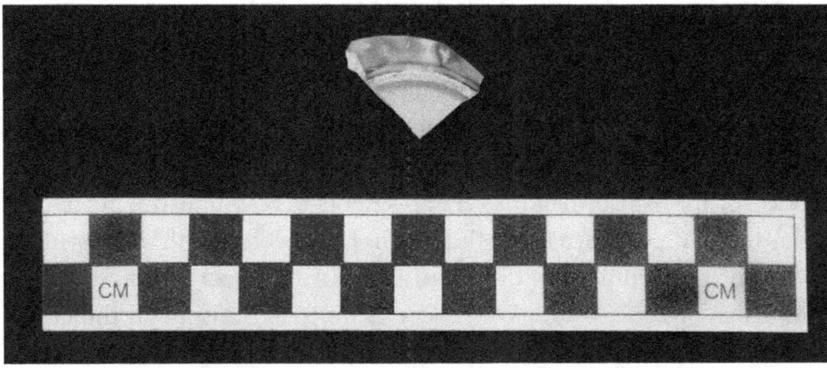

Figure 6.5. Whieldon ware *escudilla nueva* base fragment from the *tinaja* (UE 82) at La Calle Corredera, deposited in the Museo Arqueológico Municipal de Jerez de la Frontera. Image courtesy of the Museo Arqueológico Municipal de Jerez de la Frontera and the Junta de Andalucía.

While the rise in non-Spanish European ceramics in eighteenth-century St. Augustine might be attributed solely to increased trade with British colonies to the north, this explanation discounts several important factors, not the least of which is timing. In terms of official economic policy, there were no significant changes during the first half of the eighteenth century that would have spurred an influx of European goods, specifically British or British-American items, in Spanish Florida during this period. Individuals in St. Augustine had complained about the lack of regular supply shipments from Spain and Mexico for centuries; the economic policy changes designed to fix such chronic supply problems and curb illegal trade activities did not take place until 1778 (Lyon 1992; Plaza Prieto 1976:443), when the town was under British control and therefore would not have been impacted by Spanish economic policy.

Even after these official changes, the Spanish and Mexican materials available to St. Augustinians did not alter dramatically; throughout both the First and Second Spanish Periods, the residents of St. Augustine would have been able to obtain both Spanish and Mexican majolicas, which are still being produced today. These vessels virtually disappear, however, in eighteenth-century households like that of Juan de Salas.

Despite the lack of policy changes and the continuing availability of majolicas, the local economy in St. Augustine did undergo changes during the first half of the eighteenth century, although the developments had a long history. The citrus industry that coincided with increased interactions between British and Spanish territories, for example, was present in Florida by the sixteenth century but apparently not developed until 1717, when oranges became a major export product for the British colonies in the north (Harman 2004:22–24). When this industry declined in 1739 because of the War of Jenkins's Ear, residents of St. Augustine turned to piracy to obtain goods, a tactic that might have been new to the Floridian town but had been used in the Spanish Empire for over 150 years (Ewen 1991; Harman 2004:37–40; Waterbury 1983:79–81). Although the citrus industry and piracy had dramatic impacts on trade and material culture in St. Augustine, they both predate this increased relationship with the British and British Americans. They therefore can be interpreted as activities that were exploited during the first half of the eighteenth century to encourage this trade.

Consequently, it seems likely that the rise of non-Spanish European goods in St. Augustine was not necessarily the result of a sudden decision by Floridians to trade with the British in the eighteenth century to alleviate centuries of frustration with an inadequate supply system. Rather, this change in the material culture could be attributed, in St. Augustine at least, to people's attempts to stay abreast of European fashions. To do so, the residents of St. Augustine began actively seeking an increased trade relationship with British colonies and made use of available resources like citrus and opportunities like piracy to obtain fashionable goods.

Hygiene

While European fashions prevailed in public and semipublic areas of life, older Spanish vessel forms remained in use in more private areas related to personal hygiene. *Bacines* and other hygiene objects are relatively rare at the three sites. There were none at the de Salas household, and they represented between one and two percent of the *loza fina* at La Calle Corredera and the Ponce de León household.

It is perhaps worth noting that some hygiene vessels might have been made out of other materials that either did not survive or did not become part of the archaeological record. In California, for example, a copper *bacín* was found at Mission San Antonio de Padua, and both copper and pewter washbasins were mentioned in a 1777 inventory for the Mission Santa Clara de Asís (Skowronek and Thompson 2006:37–38). If metal hygiene vessels were present at Jerez and St. Augustine, they do not appear at the sites in question here.

The ceramic fragments present at these sites suggest that hygiene vessels were relatively conservative. The *bacín* form did not change until well into the nineteenth century, as indicated by examples from later deposits at La Calle Corredera and surface finds from the U.S.-Spanish Naval Station in Rota, Spain. While one tin-glazed chamber pot from the Ponce de León household appeared to take the more rounded British shape in contrast to the straight-sided *bacín* form favored by Spaniards, none of the sites yielded other fragments of chamber pots, bidets, or other non-Spanish hygiene forms.

Although the lack of change in relatively private areas is expected, it is perhaps worthwhile to consider the extent to which visitors would have actually seen these objects. Bathrooms in houses today are often decorated and updated with the same care as the rest of the house both for the sake of the residents and for any guests who might use the facilities during the course of their visit. While it can be inaccurate to assume that today's social standards are similar to those of the eighteenth century, one can reasonably conclude that eighteenth-century visitors, especially long-term houseguests, would have seen and used these more private, utilitarian, hygiene objects. Consequently, the apparent continuity of the form and the lack of new forms, such as the bidet, suggests either that they were too unimportant to upgrade or that Spaniards sought some means of maintaining some "traditional" material culture in the face of widespread changes. To pursue this topic further, we need more information from a wider range of households, including elite residences that might have been more willing and able to adopt new technology and fashions, as well as lower-status households that might have maintained more traditional lifestyles longer.

Tableware

Regardless of how often guests saw or used more private objects such as *bacines*, tableware would most likely have been seen and used by all visitors. As a result, many archaeologists, especially those working in St. Augustine and the Spanish Caribbean, have concluded that socially visible settings such as the dining table would have been among the most important areas for displaying one's social status and wealth (Deagan 1983; Ewen

1991). While clothing, jewelry, and transportation were the most publicly visible means of expressing social status and aspirations, the dining table could have been an equally important arena, as it would indicate that one was fashionable not only on the street but also in the home and knew how to properly use fashionable objects.

Consequently, one can conclude that those living at La Calle Corredera, and the Ponce de León and de Salas families, were attempting, most likely successfully, to display their knowledge of current tableware fashions and their ability to afford such goods. At all three sites, there is a marked shift in the plate and bowl forms between the seventeenth and eighteenth centuries (Tables 6.1, 6.2, and 6.3; Figures 6.6, 6.7, and 6.8). The increasing popularity of flat plates is apparent in these households, as all owned significantly more *platos nuevos* than *platos viejos*. Indeed, *platos nuevos* are the most frequently recovered forms in the individual tableware category at all three sites. In the St. Augustine sites, they are the most commonly recovered form in the whole ceramic assemblage, representing approximately 30–40 percent of the total vessels recovered, including the *loza común* ceramics studied (Ness 2015b).

One discrepancy between the sherds from the Spanish and Spanish Atlantic sites is the size of the marly, or the flat area next to the rim, on the *platos nuevos* fragments. At La Calle Corredera, many of the matching *platos nuevos* have a marly that measures 2.5 to 3 cm wide, while marlies from similar vessels from St. Augustine are typically 3.5 to 4 cm wide. While the variation is small, it is enough to make a difference when passing a plate at a dinner table. According to one early-nineteenth-century Spanish translation of a French etiquette book, the serving dishes were meant to be passed around the table, while plates remained in place (Rementeria y Fica 1837:131), which could explain the small marlies on the Spanish-made vessels. British etiquette, however, dictated that one pass his or her plate around the table so as to avoid moving the serving dishes and therefore disrupting the symmetry of the table. While it is unclear which form of etiquette St. Augustine's residents followed, their close connections with Spain and their strong desire to stay under Spanish rule after the town was ceded to the British suggest that they observed Spanish table manners, despite the fact that their dishes were designed to accommodate British etiquette.

Although Spanish potters apparently modified the way in which they copied non-Spanish European vessel forms, the fact that they began

Table 6.1. Minimum number of vessels (MNV) of tableware from La Calle Corredera

Vessel Form	MNV	%
Tableware: individual		
Escudilla nueva	16	3.26
Escudilla vieja	19	3.87
Huevero	2	0.41
Macerina	1	0.20
Plato nuevo	116	23.62
Plato nuevo pequeño	2	0.41
Plato viejo	34	6.92
Tapadera	5	1.02
Taza	64	13.03
Tableware: serving		
Ataifor	62	12.63
Bandeja	1	0.20
Salero	11	2.24
Water		
Jarrito	1	0.20
Jarro	157	31.97
Total	491	100.00

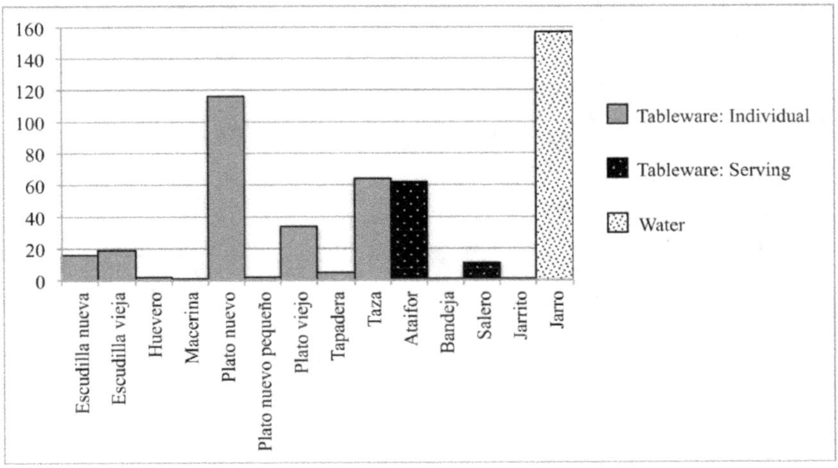

Figure 6.6. Minimum number of vessels (MNV) of tableware from La Calle Corredera. Data from Table 6.1.

Table 6.2. Minimum number of vessels (MNV) of tableware from the Ponce de León household

Vessel form	MNV	%
Tableware: individual		
Escudilla	22	15.60
Escudilla nueva	4	2.84
Jícara	1	0.71
Plato	17	12.06
Plato nuevo	50	35.46
Plato nuevo grande	4	2.84
Plato nuevo pequeño	11	7.80
Plato viejo	13	9.21
Taza	14	9.93
Tableware: serving		
Ataifor	4	2.84
Water		
Jarro	1	0.71
Total	141	100.00

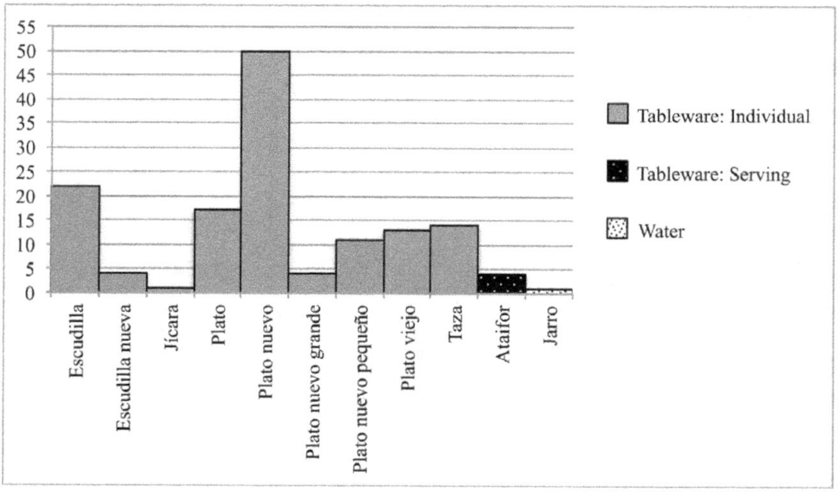

Figure 6.7. Minimum number of vessels (MNV) of tableware from the Ponce de León household. Data from Table 6.2.

Table 6.3. Minimum number of vessels (MNV) of tableware from the de Salas household

Vessel form	MNV	%
Tableware: individual		
Escudilla nueva	16	14.81
Plato nuevo	58	53.70
Plato nuevo grande	4	3.70
Plato nuevo pequeño	10	9.26
Plato viejo	3	2.78
Taza	9	8.33
Tableware: serving		
Ataifor	5	4.63
Teapot	1	0.92
Water		
Jarro	2	1.85
Total	108	100.00

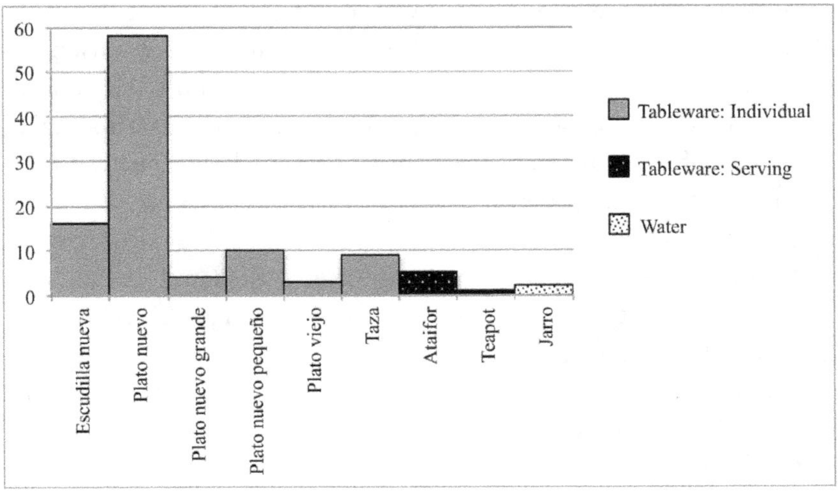

Figure 6.8. Minimum number of vessels (MNV) of tableware from the de Salas household. Data from Table 6.3.

creating new forms during this period is indicative of a dramatic change in consumer behavior. Not only did high numbers of Spaniards want the newest-style vessels, but they were also willing to change their diet in order to set a fashionable table. One cannot eat a stew, such as the traditional *olla podrida*, from a flat plate, which meant that those who were using such dishes were also eating new kinds of cuisine, possibly even "French cookery."

The eighteenth-century change in bowls is somewhat less pronounced than the change in plate forms, as indicated by the nearly identical number of *escudillas nuevas* and *escudillas viejas* at La Calle Corredera. Unlike the change in *platos*, the modification of *escudillas* was largely aesthetic and did not impact the function of the vessel. Consequently, one could have continued using *escudillas viejas* well after the new, more rounded form became popular. It seems clear from the St. Augustine sites, however, that by the middle of the eighteenth century, Spanish Americans were almost exclusively using the new form.

Tazas or *jícaras* were also present at all three sites and represented between 8 and 13 percent of the tableware assemblage across the sites. As discussed above, these vessels suggest that the families were consuming beverages like chocolate, tea, wine, or spirits, and were most likely imbibing in a socially visible setting. Because this form represents a noticeable percentage of the individual tableware at the households, one can infer the importance of the vessels and the activities they represent, such as tea or chocolate consumption.

Although all of the families owned *ataifores*, the family at La Calle Corredera used significantly more than their St. Augustinian counterparts. The discrepancy could be the result of several factors. This form is distinctly Spanish and might not have been as readily available to consumers in St. Augustine, who did not receive frequent shipments of Spanish goods. A more likely explanation, however, is that with the rise of English ceramics like white salt-glazed stoneware and creamware, the form became obsolete, and consumers began using chargers or *platos nuevos grandes*. This interpretation is supported by the presence of four *platos nuevos grandes* at both St. Augustine households and none at La Calle Corredera, which predates the rise of white salt-glazed stoneware and creamware.

Another important difference between the Spanish and Spanish American sites is the number of *jarros*. While the extreme fragility of the *paredes finas jarros* from La Calle Corredera doubtless explains their high frequency at this site, the form does appear to have been more popular in Spain than in Spanish America.

Sets

Hellman (2007) explains that despite the differences in serving tableware and water vessels, both eighteenth-century Spaniards and Spanish Americans ate from matched or semimatched dishes. At a time when all items were handmade, hand decorated, and hand washed, making and caring for matching objects required much more work than today. One flaw in production or damage caused by maintenance could ruin the effect of matching. While some eighteenth-century individuals objected to the formulaic quality of matching items, be they furniture, wall decorations, or tableware, others, as Hellman points out, found this sense of sameness pleasing. For many, it became a mark of high fashion as well as a sign of wealth, taste, and social identity. In France, matching objects were highly valued and often placed in such a way as to confer upon a space and the activities conducted within it clear attributes of symmetry and hierarchy (Hellman 2007).

Eighteenth-century Spaniards and Spanish Americans also embraced the notion of matching objects, as indicated by the archaeological remains from all three households. At La Calle Corredera, there were several "sets" of dishes that matched and represented a wide array of artistic expression, all of which were most likely made prior to Wedgwood and his well-known experiments with sets and mass production in the 1750s and 1760s. As illustrated in chapter 4, one set of two *escudillas nuevas* and two *platos nuevos* was rather crudely decorated with a simple floral motif, while several other *platos nuevos* featured the same elaborate decoration on the marly that matched other vessels attributed to potters in Triana, Sevilla. Similarly, this site included 11 *jícaras*, two *platos nuevos*, and one *escudilla nueva* that were from Alcora or a factory imitating Alcora's popular motifs. The mix of genuine and imitation Alcora ceramics suggests that the residents either had a matching set and were replacing broken pieces with copies or were attempting to supplement a smaller set with cheaper copies from other factories.

The Ponce de León and de Salas families also had matching or semimatching dishes on their tables. Both households included white salt-glazed stoneware and creamware *platos nuevos* fragments with a variety of patterns. The Ponce de Leóns, for instance, owned 20 white salt-glazed stoneware *platos nuevos*, three of which featured the "Barley" pattern, three with the "Bead and Reel" decoration, nine with the "Dot, Diaper, and Basketweave" motif, and five plain (see Florida Museum of Natural History 2012 for images and descriptions of these different patterns).

The de Salas family also had 20 white salt-glazed stoneware *platos nuevos*, four with the "Barley" pattern, one with a variation of the "Barley" pattern, one with a basketweave motif, two with the "Bead and Reel" decoration, and 12 with the "Dot, Diaper, and Basketweave" design. They also owned 13 creamware *platos nuevos*, suggesting that they had just purchased a second set of dishes around 1762 that included seven with the "Bead and Reel" pattern, one feather-edged, two with a molded decoration, and three with the "Royal" or "Queen's" pattern. This creamware set also included three *platos nuevos grandes*: one with the basketweave design, one with the "Bead and Reel" pattern, and one feather-edged.

In addition, the two sites yielded several shallow pans, probably considered *lebrillos* by Spanish American consumers, made of Staffordshire slipware. These vessels suggest that, on some level, even the less fancy, more utilitarian objects might have also matched. It is worth noting that these vessels are not the traditional *lebrillo* form, having an elongated oval shape with a round bottom and crimped rim. Given that they do not have an equivalent in Spain, and that they most likely served similar functions as *lebrillos* at these sites, I have elected to include them in this category.

Interestingly, none of the Asian porcelain fragments or *tazas* from these households appear to match other vessels at the same site. The non-matching nature of these vessels implies either that the fragments represent individual broken pieces from a set that was otherwise intact and passed down to family members or that their sets were made out of only certain types of ceramics, and other pieces were unique.

It is also perhaps worth noting that seventeenth-century deposits from other sites in St. Augustine also have "sets" of dishes. At the town's Catholic Parish School, for instance, city archaeologist Carl Halbirt found a set of four matching, seventeenth-to-eighteenth-century San Luis Polychrome *platos nuevos*. While such examples of early matching sets exist, it seems that the popularity of sets and the presence of a significant number of matching vessels do not appear until the eighteenth century.

Conclusion

The family from La Calle Corredera, the Ponce de Leóns, and the de Salas family expressed their cultural identity and their places in the Spanish Atlantic world in myriad ways. They consumed their meals on *platos nuevos*, indicating that they were willing to adopt new cooking techniques and transition away from traditional Spanish stews. The families decorated their houses with Mexican ceramics and drank chocolate and tea from specialized vessels, displaying both their wealth and their refinement to those around them. Whether they were conscious of it or not, the very ritual of preparing and consuming chocolate combined American and European elements, a perfect way of demonstrating one's participation in the Spanish Atlantic world.

The similarities among the three households become apparent when focusing on the vessel forms and using a system like COSA. The *plato nuevo* and *escudilla nueva* forms that appear on both sides of the Atlantic indicate that changes in dining practices and foodways occurred throughout the Spanish Atlantic almost simultaneously and that Spain and Spanish Florida had close ties into the 1760s. This continuing cultural connection has broad implications in terms of our understanding of the Spanish Atlantic and early modern globalization, as is discussed in the next chapter.

7

Conclusions

> Éstos fueron los versos que se pudieron leer; los demás, por estar carcomida la letra, se entregaron a un académico para que por conjeturas los declarase. Tiénese noticia que lo ha hecho, a costa de muchas vigilias y mucho trabajo, y que tiene intención de sacallos a luz, con esperanza de la tercera salida de don Quijote.
>
> [These were all the verses that could be read. The others, because of worm-eaten letters, were given to an academic to guess at their meaning. We have been told that, at the cost of many vigils and much hard work, he succeeded and intends to publish it. So we may hope for a third sally of Don Quixote. (Author's translation)]
>
> Miguel de Cervantes Saavedra, *Don Quixote*, Volume I, Chapter LII

The changing political, economic, and social climate of the eighteenth-century Spanish Atlantic forced individuals to reevaluate their sense of identity and heritage. With the flood of French influences that accompanied the Bourbon dynasty, the trade and cultural exchange with the Americas, and the increased popularity of British ceramics, Spaniards found themselves grappling with what it meant to be "Spanish" and how to express Spanish traditions.

While some activities and items, such as ceramic cooking and hygiene vessels, remained relatively unaltered during this period, other aspects of life changed dramatically. The increasing globalization of the eighteenth century is evident in new foodways, fashions, tableware, and aesthetics. As they negotiated new and old ideas, objects, and tastes, eighteenth-century individuals created a new, distinct Spanish Atlantic identity.

This developing identity was neither wholly Spanish nor wholly Spanish American but was rather an amalgamation of cultural influences from throughout the Atlantic world. As a result, it reflected the multidirectional aspects of colonialism and the global influences present in the eighteenth-century Spanish Atlantic. In the archaeological record, these influences and changes are especially apparent through ceramic vessel forms that indicate specific activities and changes to dining practices and foodways.

The family who lived at La Calle Corredera, the Ponce de Leóns, and the de Salases expressed their Spanish Atlantic identity in several similar ways. They drank imported beverages like chocolate and tea, and they ate from vessels made in Asia, England, Mexico, the Netherlands, and Spain. Their food was prepared with both local and imported ingredients and was consumed from matching, flat *platos nuevos*, indicating that they were embracing new cuisine and shifting away from more traditional Spanish stews like the *olla podrida*. The preference for matching ceramics might have been adopted from the French, but the vessels themselves were primarily made in places like England, Asia, the Netherlands, Valencia, and Triana, Sevilla.

While there were variations between the two sides of the Atlantic, such as different cookware vessels and different percentages of *ataifores* and *jarros*, the strong similarities among the sites suggest that Spain and her American territories, including relatively remote places like St. Augustine, were very closely connected even as late as the mid-eighteenth century.

Impact of these Findings

Although archaeologists have researched Spanish colonial sites in Florida and the Caribbean for decades, only two projects have adopted a trans-Atlantic perspective, and this work is the first to use this approach for the eighteenth century. Additionally, it is the first study to focus on vessel forms when studying Spanish ceramics in the Americas. As a result, this project sheds new light on the Spanish Atlantic and calls into question several existing interpretations of life in Spanish Florida, as well as foodways in both St. Augustine and Spain.

Reevaluating Interpretations of St. Augustine

Previous studies of St. Augustine suggested that wealthy individuals with Spanish ancestry would have owned and used a higher percentage of Spanish ceramics than their less well-to-do counterparts with mixed heritage (Deagan 1983:237–240; Ewen 1991). This notion, however, conflicts with the archaeological evidence from the site occupied by wealthy *peninsulare* Juan de Salas, since he appears to have owned barely any Spanish or Mexican ceramics. While his household's assemblage might be interpreted as exhibiting the crystallization of Spanish American material culture, the artifacts are similar both in terms of vessel form and ceramic type to those mentioned in descriptions of elite households in Spain and reflect the European popularity of newly-developed methods of producing hard, white-firing ceramics.

In many ways this evidence supports the idea that individuals in places like St. Augustine were conservative in maintaining "Old World Spanish affiliation as a cultural ideal" (Deagan 1983:270). In Hapsburg-era St. Augustine, it was important to demonstrate one's connections to the motherland by appearing as "Spanish" as possible and using goods from Spain and Mexico. In Bourbon-era St. Augustine, individuals were still attempting to show connections to Spain by following Spanish fashions, which then included goods and aesthetics from other parts of Europe. In other words, when one takes into account that culture and fashions in St. Augustine changed *with* culture and fashions in Spain, it appears that the individuals in Spanish America were conservative in retaining their connections to the homeland but not necessarily in keeping with traditional material culture.

Consequently, Spanish Americans appear to have been familiar with and intentionally copying trends in Spain by incorporating non-Spanish goods such as English creamwares and stonewares as well as Asian porcelains into their households. These individuals were, in a sense, reinforcing their involvement in the Spanish Atlantic. The fact that some of these ceramics were made in England and possibly obtained through illegal trade with the British colonies to the north does not necessarily indicate that the Spanish Americans in St. Augustine were abandoning their Spanish heritage. Their careful selection of some vessels over others and their apparent unwillingness to adopt forms that were not commonly used in Spain speaks to their continued desire to follow Spanish trends and to express their Spanish Atlantic identity.

It is important to remember, however, that both the Ponce de León and de Salas families represent the elite of St. Augustine, both economically and socially. For Juan de Salas, his status as a *peninsulare* could have been apparent because of his accent, relative newness to the city, or familial connections in Spain. Francisco Ponce de León, in contrast, came from an old, well-known *criollo* family. As a result, the Ponce de León and de Salas families might have expressed their Spanish heritage differently from less-affluent, *mestizo* households, such as that of the St. Augustinian de la Cruz family.

The presence of ceramics such as white salt-glazed stoneware, creamware, and Whieldon ware, at both elite and *mestizo* households, however, suggests that a wide range of St. Augustine's residents followed the prevailing fashions in Spain. As a result, it seems likely that this method of expressing Spanish Atlantic identity was common among different levels of society and that ceramic types might not have been overly significant markers of status for consumers.

Instead, it might be worthwhile to consider the vessel forms and kinds of foods that went on or in the dishes as possible indicators of status. Previous studies in St. Augustine have demonstrated that the faunal record indicates status differentiations (Reitz 1979; Reitz and Cumbaa 1983:182). If, as in Spain, different social classes in St. Augustine had access to different foodstuffs (Cervantes 1605 [1964]; Kiple 2007; Pérez Samper 2011), then perhaps food was used as a marker of status. If so, then the food that went on the plates and its preparation methods might have been equally if not more important to the early modern residents of St. Augustine than the dishes themselves. Further research is necessary, however, to determine the various ways in which less affluent individuals, like the de la Cruz family, expressed their heritage and navigated the cultural amalgamation that resulted in a Spanish Atlantic identity.

Reevaluating Dining Practices in the Spanish Atlantic

Just as the artifacts from St. Augustine require us to reevaluate long-held notions of life in the town and offer new avenues to research, the artifacts from Jerez appear to challenge documentary evidence about middle-class foodways in Spain and the extent to which French and American cultural influences infiltrated Spanish society. Although contemporary visitors to Granada wrote that the majority of people ate the stews that were staples of Spanish early modern cuisine (Townsend 1791–1792:3:94), La Calle Corredera included a significant number of plates as opposed to bowls, suggesting that the residents were eating fewer broth-based dishes and might have been following the French culinary trends that were popular in places like Madrid and Grenada. Those living in and near the site in the eighteenth century also owned and consumed American goods, as evidenced by the presence of Mexican-made Guadalajara Polychrome/ Aztec IV vessels and both *jícaras* and a *macerina* that were used to consume chocolate.

All of these vessel forms were also found at the St. Augustine sites, suggesting that foodways changed on both sides of the Atlantic almost simultaneously. While the specific individuals living at La Calle Corredera might not have set the example that was followed by Spanish Americans, it is clear from their artifacts that early modern *jerezanos* had access to a wide array of imported goods and took advantage of their connections to the global market in order to stay abreast of current fashions to the best of their ability.

Broader Implications

While this study focused on two relatively small towns within the Spanish Empire, evidence suggests that the cultural amalgamation that resulted in a Spanish Atlantic identity extended beyond these two locales. The presence of Mexican ceramics and foodstuffs like chocolate in both places indicates that those living in Mexico were also actively contributing to the Spanish Atlantic. Similarly, the presence of non-Spanish European flat plates and ceramics like creamware, whiteware, and ironstone in major Spanish ports like Havana (Ness 2015c) further supports the notion that an evolving sense of what it meant to be Spanish was not limited to Jerez and St. Augustine.

Just as individuals in Spain expressed their sense of identity differently, however, those living in the Spanish Americas might have also demonstrated their sense of Spanish traditions and identity in different ways. Studies in San Francisco, California, and Los Adaes, Texas, for example, indicate that identity for eighteenth-century Spanish Americans on the frontier was fluid and intentionally manipulated (Loren 2000; Voss 2008a), in some ways echoing those in Spain who purposely altered their clothing to signal solidarity with specific viewpoints.

Because expressions of identity and means of connecting with one's heritage can vary and depend on a number of factors, they must be taken in context. It is necessary to consider the fashions, events, and developments that occurred in Spain and other parts of the Spanish Empire while also taking into account non-Spanish influences so as to contextualize local reactions and changes. By exploring behavior through evidence such as vessel forms, texts, and dining practices, we can better understand how individuals manipulated their heritage and identity. These lines of evidence also provide clues as to what aspects they kept, modified, or rejected to suit their needs, and maybe even why they did so.

This story, however, does not stop at the borders of the Spanish Empire. Spaniards and Spanish Americans were actively trading, fighting, and working with native populations, Africans, and other non-Spanish Europeans. Although the Spanish Crown's policies for most of the eighteenth century limited official trade interactions between Spaniards and foreigners, the situation was, in reality, very fluid. In First and Second Spanish Period St. Augustine, both legal and illegal trade networks meant that Spanish Americans regularly worked with British Americans in ports like Charleston, New York, and Philadelphia. Exchanges with pirates, whether voluntary or involuntary, also meant that residents of the Florida town may have interacted with individuals from all over the world.

Likewise, the global nature of the Spanish Empire in the eighteenth century meant that those living under the Spanish Crown may have had connections far beyond the empire's official boundaries. In St. Augustine there were Spaniards, Englishmen, Germans, Frenchmen, Italians, Greeks, Swiss, Americans, Minorcans, Canary Islanders, Scots, Irish, Native Americans, and Africans—any number of whom might have had connections beyond the Floridian town and the Spanish Empire. The population in Spain also was diverse, as there were frequent visitors as well as permanent residents from both the Americas and other European countries throughout the Iberian Peninsula, including towns like Cádiz and Jerez. Consequently, whether communicating with nonlocal family members and friends or engaging in trade, those living in places like Jerez and St. Augustine frequently had opportunities to exchange goods and ideas with people from multiple countries and empires.

This idea of multidirectonal culture exchange guided this study to an extent. It explored the impacts that Spain and the Spanish Americas had on each other in the eighteenth century and how people in these areas responded to outside influences like French fashions and British items. Individuals on both sides of the Atlantic attempted to reinterpret and incorporate new ideas and objects into a developing Spanish Atlantic identity that reflected both traditional Spanish culture and the increasing globalization of the period, a trend that is especially apparent in dining practices. While this approach is valid for numerous imperial settings, it could also be useful for studying the interactions between empires, such as the Spanish and British Empires. There are many more directions to explore and many stories left to tell about the early modern Atlantic world.

References Cited

Archival Documents and Collections

Archivo General de Indias (AGI)

1787/1817 Cargos y datas. ES.41091.AGI/27.61//CUBA,2322B. Accessed through the *Portal de Archivos Españoles*, http://pares.mcu.es/ParesBusquedas/servlets/Control_servlet?accion=3&txt_id_desc_ud=61457&fromagenda=S, June 5, 2014.

City of St. Augustine

2014 *Code of the City of St. Augustine, Florida*. Ordinance No. 13–36.

East Florida Papers (EFP)

1794 Sublieutenant Rocco Leonardy to Governor of Florida, 12 July, Reel 78, Bundle 184B15. Library of Congress Microfilm Collection.

1799 Clemente Salas to Governor of Florida, 17 July, Reel 79, Bundle 188F15. Library of Congress Microfilm Collection.

1801 Tax Records for the Bario de Iglecia, St. Augustine, 1 April, Reel 147, Bundle 322. Library of Congress Microfilm Collection.

Florida Master Site Files

n.d. MC 54, Block 4, Lot 1, 2, 3, F.19. Manuscript on file at the St. Augustine Historical Society, St. Augustine.

2003 8IR00024: Higgs Point Site. Manuscript on file at the Florida Department of State, Division of Historical Resources.

Florida Museum of Natural History

n.d. *S.A. 36–4 Collection Summary*. Manuscript on file, Department of Anthropology, Florida Museum of Natural History, Gainesville.

2012 Historical Archaeology Digital Type Collection. Electronic document, http://www.flmnh.ufl.edu/histarch/gallery_types/, accessed November 1, 2015.

St. Augustine Historical Society Research Library

1759 Letter from Juan Cajigal de la Vega to Julian de Arriaga, April 11, 1759. AGI 87-3-13/20I. Accessed through the Stetson Collection of Microfilms at the St. Augustine Historical Society.

1766 St. Augustine History MC 63, 1766 Box 5, File 1, Puente. Ref. of Former St. Augustine Families in Havana (AGI 87-5-3). Manuscript on file, St. Augustine, Florida.

University of Florida

2014 University of Florida Digital Collections. Electronic document, http://ufdc.ufl.edu/usach, accessed November 1, 2015.

Secondary Sources

Albala, Ken

2006 *Cooking in Europe, 1250–1650*. Greenwood Press, Westport, Connecticut.

Altamiras, Juan

1758 *Nuevo arte de cocina sacado de la escuela de la experiencia económica su autor*. Juan de Bezáres, Barcelona.

Amores Carredano, Fernando de, and Nieves Chisvert Jiménez

1993 Tipología de la cerámica común bajomedieval y moderna sevillana (ss. XV–XVIII): I. La loza quebrada de relleno de bóvedas. *SPAL* 2:269–325.

Andalucía Dirección General de Bienes Culturales

1985–2006 *Anuario arqueologico de Andalucía*. Direccion General de Bienes Culturales, Sevilla.

Appler, Douglas R.

2012 Municipal Archaeology Programs and the Creation of Community Amenities. *Public Historian* 34(3):40–67.

Armitage, David, and Michael J. Braddick, eds.

2002 *The British Atlantic World, 1500–1800*. Palgrave Macmillan, London.

Arnade, Charles W.

1961 The Architecture of Spanish St. Augustine. *The Americas* 18(2):149–186.

Aroca Vicenti, Fernando

2002 *Arquitectura y urbanismo en el Jerez del siglo XVIII*. Digital Impresión, Jerez de la Frontera, Spain.

Bas, Laura R., and Amanda Wunder

2009 The Veiled Ladies of the Early Modern Spanish World: Seduction and Scandal in Sevilla, Madrid, and Lima. *Hispanic Review* 77(1):97–144.

Bayliss, Robert

2006 What *Don Quixote* Means (Today). *Comparative Literature Studies* 43(4): 382–397.

Beaudry, Mary C.

2008 "Above Vulgar Economy": The Intersection of Historical Archaeology and Microhistory in Writing Archaeological Biographies of Two New England Merchants. In *Small Worlds: Method, Meaning, and Narrative in Microhistory*, edited by James F. Brooks, Christopher R. N. Decorse, and John Walton, pp. 173–198. School for Advanced Research Press, Santa Fe.

Beaudry, Mary C., Janet Long, Henry M. Miller, Fraser D. Neiman, and Garry Wheeler Stone

1983 A Vessel Typology for Early Chesapeake Ceramics: The Potomac Typological System. *Historical Archaeology* 17(1):18–43.

Benedict, Ruth

1934 *Patterns of Culture*. Houghton Mifflin, Boston.

Binford, Lewis R.

1962 Archaeology as Anthropology. *American Antiquity* 28(2):217–225.

Boyd-Bowman, Peter

1976 Patterns of Spanish Emigration to the Indies until 1600. *Hispanic American Historical Review* 56(4):580–604.

Branch, Taylor

1998 *Pillar of Fire: America in the King Years, 1963–1965*. Simon and Schuster, New York.

Buker, George E.

1983 The Americanization of St. Augustine: 1821–1865. In *The Oldest City, St. Augustine: Saga of Survival*, edited by Jean Parker Waterbury, pp. 151–180. St. Augustine Historical Society, St. Augustine.

Bushnell, Amy

1983 The Noble and Loyal City: 1565–1688. In *The Oldest City, St. Augustine: Saga of Survival*, edited by Jean Parker Waterbury, pp. 27–56. St. Augustine Historical Society, St. Augustine.

Cadalso, Joseph

1796 *Cartas marruecas del Coronel*. La Imprenta de Piferrer, Barcelona.

Cadenhead, Ivie E. Jr.,

1960 Some Mining Operations of Cortes in Tehuantepec, 1538–1547. *The Americas* 16(3):283–287.

Cannadine, David

2001 *Ornamentalism: How the British Saw Their Empire*. Oxford University Press, Oxford.

Caro, Diego (editor)

1999 *Historia de Jerez de la Frontera: Tomo I. De los orígenes a la época medieval*. Diputación de Cádiz, Cádiz.

Carrera, Magali M.

1998 Locating Race in Late Colonial Mexico. *Art Journal* 57 (3):36–45.

2003 *Imagining Identity in New Spain: Race Lineage, and the Colonial Body in Portraiture and Casta Paintings*. University of Texas Press, Austin.

Carroll, Lynda

1999 Communities and Other Social Actors: Rethinking Commodities and Consumption in Global Historical Archaeology. *International Journal of Historical Archaeology* 3(3):131–136.

Castro, Adolfo de

1845 *Historia de Xerez de la Frontera*. La Sociedad de la Revista Medica, Cádiz.

Cervantes de Saavedra, Miguel
1605 *El ingenioso hidalgo don Quijote de la Mancha*. Francisco Murcuia de la Llana.
1964 *Don Quixote*. Translated by Walter Starkie. Macmillan, London.
Coe, Sophie D., and Michael D. Coe
2013 *The True History of Chocolate*. Thames and Hudson, London.
Covarrubias, Sebastián de
1611 *Tresoro de la lengua castellana o española*. Luis Sánchez, Madrid.
Crown, Patricia L.
2012 Chocolate Consumption and Cuisine: From Chaco to Colonial New Mexico. *El Palacio* 4:36–43.
Cusick, James Gregory
1993 *Ethnic Groups and Class in an Emerging Market Economy: Spaniards and Minorcans in Late Colonial St. Augustine*. Ph.D. dissertation, Department of Anthropology, University of Florida. University Microfilms, Ann Arbor, Michigan.
2000 Creolization and the Borderlands. *Historical Archaeology* 34(3):46–55.
d'Aulnoy, Marie-Catherine
1692 *Relation du voyage d'Espagne*. Chez Henri van Bulderen, La Haye.
Deagan, Kathleen
1975 Architecture, Archaeology and History in Spanish St. Augustine: New Data from Sealed Contexts. In *The Conference on Historical Sites Archaeology Papers, Vol. 9*, edited by Stanley South, pp. 13–29. University of South Carolina, Columbia.
1983 *Spanish St. Augustine: The Archaeology of a Colonial Creole Community*. Academic Press, New York.
1987 *Artifacts of the Spanish Colonies of Florida and the Caribbean, 1500–1800: 1. Ceramics, Glassware, and Beads*. Smithsonian Institution Press, Washington, D.C.
Dennis, George
1839 *A Summer in Andalucía*, Vols. 1 and 2. Richard Bentley, London.
Díaz, Lorenzo
2005 *Ilustrados y románticos: Cocina y sociedad en España (siglos XVIII y XIX)*. Alianza Editorial, Madrid.
Dillon, Sir John Talbott
1781 *Travels through Spain*. S. Price, W. and H. Whitestone, T. Walker, W. Gilbert, C. Jenkin, W. Hallhead, L. White, C. Talbot, and P. Byrne, Dublin.
Donahue, Darcy
2004 Dressing Up and Dressing Down: Clothing and Class Identity in the *Novelas ejemplares*. *Cervantes: Bulletin of the Cervantes Society of America* 24(1):105–118.
Elliott, J. H.
1963 *Imperial Spain: 1469–1716*. Penguin Books, New York.

Esteve Guerrero, Manuel
1962 El casco urbano de Jerez de la Frontera. *Publicaciones del Centro de Estudios Historicos Jerezanos* 18:7–18.

Ewen, Charles
1991 *From Spaniard to Creole: The Archaeology of Cultural Formation at Puerto Real, Haiti.* University of Alabama Press, Tuscaloosa.

Fairbanks, Charles H. 1973 The Cultural Significance of Spanish Ceramics. In *Ceramics in America*, edited by Ian M. G. Quimby, pp. 141–174. University Press of Virginia, Charlottesville.

1975 Spanish Artifacts at the Fortress of Louisbourg, Cape Breton Island. In *The Conference on Historic Site Archaeology Papers, 1974*, edited by Stanley South, pp. 30–59. Institute of Archaeology and Anthropology, University of South Carolina, Columbia.

Fattacciu, Irene
2009 Cacao: From an Exotic Curiosity to Spanish Commodity, the Diffusion of New Patterns of Consumption in Eighteenth-Century Spain. *Food and History* 7(1):53–78.

2011 Gremios y evolución de las pautas de consume en el siglo XVIII: La industria artisanal del chocolate. In *Comprar, vender y consumer: Nuevas aportaciones a la historia del consume en la España moderna*, edited by Daniel Muñoz Navarro, pp. 153–171. Universitat de Valéncia, Valéncia.

Friede, Juan
1951 The Catálogo de Pasajeros and Spanish Emigration to America in 1550. *Hispanic American Historical Review* 13(2):333–348.

Garcia Iñañez, Javier, Robert J. Speakman, Jaume Buxeda i Garrigós, and Michael D. Glascock
2008 Chemical Characterization of Majolica from 14th–18th Century Production Centers on the Iberian Peninsula: A Preliminary Neutron Activation Study. *Journal of Archaeological Science* 35:425–440.

2009a Chemical Characterization of Tin-Lead Glazed Pottery from the Iberian Peninsula and the Canary Islands: Initial Steps toward Better Understanding of Spanish Colonial Pottery in the Americas. *Archaeometry* 51(4):546–567.

2009b Archaeometric Characterization of Renaissance Tin-Lead Glazed Pottery from Talavera de la Reina, Puente del Arzobispo and Seville (Spain). In *Proceedings of the 36th International Symposium on Archaeometry*, edited by Jean-François Moreau, Réginald Auger, Jacques Chabot, and Anja Herzog, pp. 279–287. Université Laval, Quebec.

Gasco, Janine
1992 Documentary and Archaeological Evidence for Household Differentiation in Colonial Socunusco, New Spain. In *Text-Aided Archaeology*, edited by Barbara J. Little, pp. 83–94. CRC Press, Boca Raton, Florida.

Gibson, Charles
1966 *Spain in America.* Harper and Row, New York.

Glas, George
1764 *The History of the Discovery and Conquest of the Canary Islands: Translated from a Spanish Manuscript, lately found in the Island of Palma with an Enquiry into the Origin of the Ancient Inhabitants. To which is added, A Description of the Canary Islands, Including the Modern History of the Inhabitants, and an Account of their Manners, Customs, Trade, &c.* R. and J. Dudley and T. Durham, London.

Glasco, Janine L.
2005 Spanish Colonialism and Processes of Social Change in Mesoamerica. In *The Archaeology of Colonial Encounters: Comparative Perspectives,* edited by Gil Stein, pp. 69–108, School of American Research Press, Oxford.

Goggin, John M.
1960 *The Spanish Olive Jar: An Introductory Study.* Yale University Publications in Anthropology, No. 62. Yale University, New Haven, Connecticut.
1968 *Spanish Majolica in the New World: Types of the Sixteenth to Eighteenth Centuries.* Yale University Publications In Anthropology, No. 72. Yale University, New Haven, Connecticut.

González Beltrán, Jesús, and José Luis Pereira Iglesias
1999 *Prima parte: Jerez de la Frontera en la edad moderna.* In *El Jerez moderno y contemporáneo,* edited by Diego Caro Cancela, pp. 13–194. Diputación de Cádiz, Servico de Publicaciones, Cádiz.

González Martí, Manuel
1936 *Cerámica española.* Talleres Gráficos Ibero-Americanos, S.A., Barcelona.

González Rodríguez, Rosalía, Francisco Barrionuevo Contreras, and Laureano Aguilar Moya
1997 *Museo Arqueológico Municipal Jerez de la Frontera.* Museo Arqueológico Municipal, Jerez de la Frontera, Spain.

González Rodríguez, Rosalía, Laureano Moya Aguilar, Domingo Martín Mochales, Barrionuevo Francisco Contreras, and Manuel Moreno Collado
2008 *Carta arqueológica municipal de Jerez. 1: El núcleo urbano.* Consejería de Cultura, Sevilla.

Granado, Diego
1614 *Libro del arte de cozina, en el qual se contiene el modo de guisar de comer en qualquier tiempo, assi de carne, como de pescado, para sanos y enfermos, y conualecientes: Assi de pasteles, tortas, y salsas, como de conseruas a la usança española, italiana, y tudesca de nuestros tiempos.* Luys Manescal, Lerida.

Griffin, Patricia C.
1983 The Spanish Return: The People-Mix Period: 1784–1821. In *The Oldest City, St. Augustine: Saga of Survival,* edited by Jean Parker Waterbury, pp. 125–150. St. Augustine Historical Society, St. Augustine.

Gutiérrez López, José María
2001 Intervención arqueológica de urgencia en la Batería Duque de Nájera (Rota, Cádiz), 1996–1997. In *Anuario arqueologico de Andalucía 1997: Actividades*

de urgencia, edited by Andalucía Dirección General de Bienes Culturales, pp. 116–127. Consejería de Cultura de Junta de Andalucía, Sevilla.

Halbirt, Carl D.

2003 Synopsis of Archaeological Excavations at the Monson Motor Lodge, St. Augustine Florida, and a Description of Three Trash Deposits. Electronic document, http://ufdc.ufl.edu//USACH00405/00002, accessed November 24, 2013.

Haring, Clarence Henry

1915 American Gold and Silver Production in the First Half of the Sixteenth Century. *Quarterly Journal of Economics* 29(3):433–479.

1918 *Trade and Navigation between Spain and the Indies in the Time of the Hapsburgs*. Harvard University Press, Cambridge, Massachusetts.

Harrington, J. C., Albert C. Manucy, and John M. Goggin

1955 Archeological Excavations in the Courtyard of Castillo de San Marcos, St. Augustine, Florida. *Florida Historical Quarterly* 34(2):101–141.

Harman, Joyce Elizabeth

2004 *Trade and privateering in Spanish Florida, 1732–1763*. University of Alabama Press, Tuscaloosa.

Hellman, Mimi

2007 The Joy of Sets: The Uses of Seriality in the French Interior. In *Furnishing the Eighteenth Century: What Furniture Can Tell Us about the European and American Past*, edited by Dena Goodman and Kathryn Norberg, pp. 129–153. Routledge, New York.

Henry, Susan L.

1991 Consumers, Commodities, and Choices: A General Model of Consumer Behavior. *Historical Archaeology* 25(2):3–14.

Herr, Richard.

1958 *The Eighteenth-Century Revolution in Spain*. Princeton University Press, Princeton, New Jersey.

Isbert, Francesc, and Christina Andreu, editors

2006 *Sa Raval des Castell: La història d'un poble a través del material arqueològic*. Museu de Menorca, Mahon, Spain.

James, Stephen R.

1988 A Reassessment of the Chronological and Typological Framework of the Spanish Olive Jar. *Historical Archaeology* 22(1):43–66.

Jones, Julie

2002 Gold of the Indies. *Metropolitan Museum of Art Bulletin* 59(4):3–4.

Kellogg, Susan

2000 Depicting *Mestizaje*: Gendered Images of Ethnorace in Colonial Mexican Texts. *Journal of Women's History* 12(3):69–92.

Kiddle, Lawrence Bayard

1944 *The Spanish Word* Jícara, *a Word History, with an Appendix of the Manufacture of Jícaras in Olinalá, Guerrero*. Middle American Research Institute, Tulane University, New Orleans.

Kiple, Kenneth F.
2007 *A Movable Feast: Ten Millennia of Food Globalization*. Cambridge University Press, Cambridge.

Krause, William Craig
1993 Casticismo and the Music of Federico Moreno Torroba. *Revista de Musicología* 16(6):3235–3241.

Kuznesof, Elizabeth Anne
1995 Ethnic and Gender Influences on "Spanish" Creole Society in Colonial Spanish America. *Colonial Latin American Review* 4(1):153–176.

Labat, Jean-Baptiste
1730 *Voyages du P. Labat de l'Ordre des FF Precheurs, en Espagne et en Italie, Tome I*. Jean-Baptiste Delespine, Paris.

Landers, Jane
1999 *Black Society in Spanish Florida*. University of Illinois Press, Urbana.

Lavedan, Don Antonio
1796 *Tratado de los usos, abusus, propiedades y vertudes del tabaco, café, té y chocolate*. Imprenta Real, Madrid.

Lister, Florence, and Robert Lister
1974 Maiolica in Colonial Spanish America. *Historical Archaeology* 8(1):17–52.
1976 *A Descriptive Dictionary for 500 Years of Spanish-Tradition Ceramics (13th through 18th Centuries)*. Society for Historical Archaeology, Glassboro, New Jersey.
1987 *Andalusian Ceramics in Spain and New Spain*. University of Arizona Press, Tucson.

Lockhart, James
1976 Letters to People in Spain. In *First Images of Americas: The Impact of the New World on the Old*, Vol. II, edited by Fredi Chiappelli, Michael J. B. Allen, and Robert L. Benson, pp. 783–796. University of California Press, Los Angeles.

Logan, Patricia Ann
1977 *The San José y Las Animas: An Analysis of the Ceramic Collection*. M.A. thesis, Department of Anthropology, Florida State University, Tallahassee.

Longacre, William A.
2010 Archaeology as Anthropology Revisited. *Journal of Archaeological Method and Theory* 17(2):81–100.

Longhurst, Brian
2007 *Cultural Change and Ordinary Life*. Open University Press, New York.

Longoria, Linda Dale
2007 *Chinese Export Porcelain in the Missions and Presidios of Eighteenth Century Spanish Texas*. M.A. thesis, Department of Anthropology, University of Texas, San Antonio.

López Rosendo, Ester
2005a *Arqueología de época moderna en una vivienda burguesa de Jerez de la Fron-*

tera (Cádiz). Intervención arqueológica preventiva en la C/ Corredera No. 39. In *Anuario arqueologico de Andalucía*, edited by Andalucía Dirección General de Bienes Culturales, pp. 280–294. Direccion General de Bienes Culturales, Sevilla.

2005b Inventario de materiales. Manuscript on file, Museo Arqueológico Municipal, Jerez de la Frontera, Spain.

2005c *Memoria final de la Intervención Arqueológica Preventiva desorrollada en la C/Corredera no. 39 de Jerez de la Frontera (Cádiz)*. Manusript on file, Museo Arqueológico Municipal, Jerez de la Frontera, Spain.

Loren, Diana DiPaolo

2000 The Intersections of Colonial Policy and Colonial Practice: Creolization on the Eighteenth-Century Louisiana/Texas Frontier. *Historical Archaeology* 34(3):85–98.

2007 Corporeal Concerns: Eighteenth-Century Casta Paintings and Colonial Bodies in Spanish Texas. *Historical Archaeology* 41(1):23–36.

2008 *In Contact: Bodies and Spaces in the Sixteenth- and Seventeenth-Century Eastern Woodlands*. AltaMira Press, Lanham, Maryland.

Lyon, Eugene

1992 Richer Than We Thought: The Material Culture of Sixteenth-Century St. Augustine. *El Escribano*, 29:1–117.

Machuca, Paulina

2012 De porcelanas chinas y otros menesteres: Cultural material de origen asiático en Colima, siglos XVI–XVII. Relaciones 33(131):77–134.

Manucy, Albert

1978 *The Houses of St. Augustine: 1565–1821*. St. Augustine Historical Society, St. Augustine.

Martínez Bermejo, Saúl

2008 Beyond Luxury: Sumptuary Legislation in 17th-Century Castile. In *Making, Using and Resisting the Law in European History*, edited by Günther Lottes, Eero Matijainen, and Jón Vidar Sigurdsson, pp. 93–108. PLUS-Pisa University Press, Pisa.

Martínez Montiño, Francisco

1611 *Arte de cozina, pasteleria, vizcocheria, y conserueria*. Luis Sanchez, Madrid.

McEwan, Bonnie G.

1989 *An Archaeological Perspective of Sixteenth-Century Spanish Life in the Old World and the Americas*. Ph.D. dissertation, Department of Anthropology, University of Florida. University Microfilms International (UMI), Ann Arbor, Michigan.

McNeil, Cameron L.

2009 Introduction: The Biology, Antiquity, and Modern Uses of the Chocolate Tree (*Theobroma cacao* L.). In *Chocolate in Mesoamerica: A Cultural History of Cacao*, edited by C. L. McNeil, pp. 1–28. University Press of Florida, Gainesville.

Mills, Donna Rachal

1992 *Florida's First Families: Translated Abstracts of Pre-1821 Spanish Censuses.* Mills Historical Press, Naples, Florida.

Minsheu, John

1617 *Vocabularium Hispanicum Latinum et Anglicum coppiossisimum, cum nonullis vocum millibus locupletatum, ac cum Linguae Hispanica Etymologijs.* Joanum Browne, London.

Mörner, Magnus

1976 Spanish Migration to the New World prior to 1800: A Report on the State of Research. In *First Images of Americas: The Impact of the New World on the Old*, Vol. II, edited by Fredi Chiappelli, Michael J. B. Allen, and Robert L. Benson, pp. 723–736. University of California Press, Los Angeles.

1995 Spanish Historians on Migration to America during the Colonial Period. *Latin American Research Review* 30(2):251–267.

Muller, Priscilla E.

1972 *Jewels in Spain, 1500–1800.* Hispanic Society of America, New York.

Museo Arqueologica Nacional

2014 *Edad moderna.* Exhibition, Madrid.

Myers, J. Emlen, Carredano F. de Amores, Jacqueline S. Olin, and A. Pleguezuelo Hernandez

1992 Compositional Identification of Seville Majolica at Overseas Sites. *Historical Archaeology* 26(1):131–147.

Ness, Kathryn L.

2015a Classification Systems with a Plot: Vessel Forms and Ceramic Typologies in the Spanish Atlantic. *International Journal of Historical Archaeology* 19(2):309–333.

2015b *Developing a Spanish-Atlantic Identity: An Archaeological Investigation of Domestic Ceramics and Dining in 18th-Century Spain and Spanish Florida.* Ph.D. dissertation, Department of Archaeology, Boston University, Boston.

2015c Globalization and Luxury Ceramics of the 18th and 19th-Century Spanish-Atlantic World. In *Global Pottery: Historical Archaeology and Archaeometry for Societies in Contact*, edited by J. Buxeda i Garrigós, M. Madrid i Fernández, and Javier G. Iñañez, pp. 263–272. Archaeopress, BAR International Series 2761, Oxford.

Norton, Marcy

2006 Tasting Empire: Chocolate and the European Internalization of Mesoamerican Aesthetics. *American Historical Review* 111(3):660–691.

Noyes, Dorothy

1998 La Maja Vestida: Dress as Resistance to Enlightenment in Late-18th-Century Madrid. *Journal of American Folklore* 111(440):197–217.

Olin, Jacqueline S., and M. James Blackman

1989 Compositional Classification of Mexican Majolica Ceramics of the Spanish Colonial Period. In *Archaeological Chemistry IV*, edited by Ralph O. Allen, pp. 87–112. American Chemical Society, Washington, D.C.

Olin, Jacquline S., Garmon Harbottle, and Edward Sayre
1978 Elemental Compositions of Spanish and Spanish Colonial Majolica Ceramics in the Identification of Provenience. In *Archaeological Chemistry, II: Proceedings*, edited by Giles F. Carter, pp. 200–229. American Chemistry Society, Washington, D.C.
Otto, John Solomon, and Russell Lamar Lewis Jr.
1974 A Formal and Functional Analysis of San Marcos Pottery from Site SA 16–23 St. Augustine, Florida. Bureau of Historic Sites and Properties, Bulletin No. 4. Florida Department of State, Tallahassee.
Parker, Susan B.
1999 *The Second Century of Settlement in Spanish St. Augustine, 1670–1763*. Ph.D. dissertation, Department of History, University of Florida. University Microfilms International (UMI), Ann Arbor, Michigan.
Percival, Richard
1591 *Bibliotecae Hispanicae pars altera. Containing a Dictionarie in Spanish, English and Latine.* John Jackson and Richard Watkins, London.
Pérez Samper, María de los Angeles
2009 La alimentación cotidiana en la Española del siglo XVIII. In *La vida cotidiana en la España del siglo XVIII*, edited by Manuel-Reyes García Hurtado, pp. 11–56. Sílex, Madrid.
2011 *Mesas y cocinas en la España del siglo XVIII*. Tria, Gijón, Spain.
Perissinotto, Giorgio (editor)
1998 *Documenting Everyday Life in Early Spanish California: The Santa Barbara Presidio Memorias y Facturas, 1779–1810*. Santa Barbara Trust for Historic Preservation, Santa Barbara.
Piernas Hurtado, José Manuel
1907 *La Casa de la Contratacion de las Indias*. V. Suárez, Madrid.
Pike, Ruth
1972 *Aristocrats and Traders: Sevillian Society in the Sixteenth Century*. Cornell University Press, Ithaca.
Plaza Prieto, Juan
1976 *Estructura económica de España en el siglo XVIII*. Confederación Española de Cajas de Ahorros, Madrid.
Pleguezuelo, Alfonso
2011 *Lozas y azulejos de Triana: Colección Carranza*. Ayuntamiento de Sevilla, Servicio de Publicacions, Sevilla.
Pountain, Christopher J.
2000 *A History of the Spanish Language through Texts*. Routledge, London.
Ramos Palencia, Fernando
2010 *Pautas de consumo y mercado en Castilla: 1750–1850; Economía familiar en Palencia al final del Antiguo Régimen*. Silex, Madrid.
2011 *Notas metodológicas sobre la utilización de los inventarios post-mortem: Clasificación de bienes de consumo, bases de datos e impacto de créditos y duedas, 1700–1850*. In *Comprar, vender y consumir: Nuevas aportaciones a la historia*

del consumo en la España moderna, edited by Daniel Muñoz Navarrao, pp. 21–52. Universitat de Valéncia, Valéncia.

Ray, Anthony
2000 *Spanish Pottery, 1248–1898: With a Catalogue of the Collection of the Victoria and Albert Museum*. V&A Publications, London.

Real Academia Española
1726–1739 *Diccionario de autoridades*. Nuevo Tesoro Lexicográfico de la Lengua Española, Madrid. Electronic document, http://ntlle.rae.es/ntlle/SrvltGUILoginNtlle, accessed October 2, 2015.
1770 *Diccionario de autoridades (A–B)*, 2nd ed. Nuevo Tesoro Lexicográfico de la Lengua Española, Madrid. Electronic document, http://ntlle.rae.es/ntlle/SrvltGUILoginNtlle, accessed October 2, 2015.
1780 *Diccionario de la lengua castillana*. Nuevo Tesoro Lexicográfico de la Lengua Española, Madrid. Electronic document, http://ntlle.rae.es/ntlle/SrvltGUILoginNtlle, accessed October 2, 2015.
1803 *Diccionario de la lengua castillana*. 4th ed. Nuevo Tesoro Lexicográfico de la Lengua Española, Madrid. Electronic document, http://ntlle.rae.es/ntlle/SrvltGUILoginNtlle, accessed October 2, 2015.

Reitz, Elizabeth J.
1979 *Spanish and British Subsistence Strategies at St. Augustine, Florida, and Frederica Georgia, between 1565 and 1763*. Ph.D. dissertation, Department of Anthropology, University of Florida. University Microfilms International (UMI), Ann Arbor, Michigan.

Reitz, Elizabeth J., and Stephen L. Cumbaa
1983 Diet and Foodways of Eighteenth-Century Spanish St. Augustine. In *Spanish St. Augustine: The Archaeology of a Colonial Creole Community*, edited by Kathleen Deagan, pp. 151–186. Academic Press, New York.

Reitz, Elizabeth J., and C. Margaret Scarry
1985 *Reconstructing Historic Subsistence with an Example from Sixteenth-Century Spanish Florida*. Special Publication Series, 3. Society for Historical Archaeology, Rockville, Maryland.

Rementeria y Fica, Don Mariano de
1837 *El hombre fino al gusto del dia, ó Manual complete de urbanidad, cortesia y buen tono, con las reglas, aplicaciones y ejemplos del Arte de presentarse y conducirse en toda clase de reunions, visitas, etc.; En el que se enseña la etiqueta y ceremonial que la sensatez y la costumbres han establecido; Con la guia del tocador y un tratado del Arte cisoria. Traduccion del Frances al Castellano*. Colegio de Sordo-Mudos, Madrid.

Rivas Pérez, Jorge F.
2013 Domestic Display in the Spanish Overseas Territories. In *Behind Closed Doors: Art in the Spanish American Home, 1492–1898*, edited by Richard Aste, pp. 49–104. Brooklyn Museum, Brooklyn, New York.

Rosal, Francisco del
1611 *Origen y etymología de todos los vocablos originales de la lengua castellana*.

Manuscript on file Biblioteca Nacional de Madrid, Ms. 6929. Electronic document, http://ntlle.rae.es/ntlle/SrvltGUILoginNtlle, accessed October 2, 2015.

Roscoe, Thomas

1836 *The Tourist in Spain.* Robert Jennings, London.

Rothschild, Nan A.

2006 Colonialism, Material Culture, and Identity in the Rio Grande and Hudson River Valleys. *International Journal of Historical Archaeology* 10(1):73–108.

Rouse, Irving

1951 *A Survey of Indian River Archeology, Florida.* Yale University Publications in Anthropology No. 44. Yale University Press, New Haven, Connecticut.

1960 The Classification of Artifacts in Archaeology. *American Antiquity* 25(3):313–323.

Ruiz Gil, José Antonio

1999 *Arqueología en la Bahía de Cádiz durante la edad moderna.* Ph.D. dissertation, Departamento de Historia de la Antigüedad, del Arte, Geografía Física y Antropología, Universidad de Huelva, Huelva, Spain.

St. Augustine Historical Society

1955–2014 *El Escribano.* St. Augustine Historical Society, St. Augustine.

Santos, Agapita Jurado

2012 De la risa a carajadas a mal ejemplo quijotesco en la novela del XVIII. Don Quixote de la Manchuela. *Lingue e letterature d'oriente e d'occidente* 1(1):277–286.

Schafer, Daniel L.

1983 ". . . not so gay a Town in America as this . . .": 1763–1784. In *The Oldest City, St. Augustine: Saga of Survival,* edited by Jean Parker Waterbury, pp. 91–124. St. Augustine Historical Society, St. Augustine.

Scott, Elizabeth M.

1997 "A Little Gravy in the Dish and Onions in a Tea Cup": What Cookbooks Reveal about Material Culture. *International Journal of Historical Archaeology* 1(2):131–155.

Seed, Patricia

1982 Social Dimensions of Race: Mexico City, 1753. *Hispanic American Historical Review* 62(4):569–606.

Shephard, Steven J.

1983 The Spanish *Criollo* Majority in Colonial St. Augustine. In *Spanish St. Augustine: The Archaeology of a Colonial Creole Community,* edited by Kathleen Deagan, pp. 65–97. Academic Press, New York.

Skowronek, Russell K.

1984a *Trade Patterns of Eighteenth-Century Frontier New Spain: The 1733 Flota and St. Augustine.* South Carolina Institute of Archaeology and Anthropology, University of South Carolina, Columbia.

1984b Ceramics and Commerce: The 1554 *Flota* Revisited. *Historical Archaeology* 21(2):101–111.

1991 Paintings and Potsherds. *Archaeology* 44(6):52–59.

1992 Empire and Ceramics: The Changing Role of Illicit Trade in Spanish America. *Historical Archaeology* 26(1):109–118.

Skowronek, Russell K., M. James Blackman, and Ronald J. Bishop

2014 *Ceramic Production in Early Hispanic California: Craft, Economy, and Trade on the Frontier of New Spain.* University Press of Florida, Gainesville.

Skowronek, Russell K., with Elizabeth Thompson

2006 *Situating Mission Santa Clara de Asís: 1776–1851, Documentary and Material Evidence of Life on the Alta California Frontier: A Timeline.* Academy of American Franciscan History, Berkeley, California.

South, Stanley, Russell K. Skowronek, and Richard E. Johnson

1988 *Spanish Artifacts from Santa Elena.* Anthropological Studies, Vol. 7. University of South Carolina, South Carolina Institute of Archaeology and Anthropology, Columbia.

Stevens, John (trans.)

1702 *The Spanish Rule of Trade to the West-Indies.* Samuel Crouch, London.

Steward, Julian Haynes

1976 *Theory of Culture Change: The Methodology of Multilinear Evolution.* University of Illinois Press, Urbana.

Tax, Sol

1951 Selective Culture Change. *American Economic Review* 41(2):315–320.

TePaske, John Jay

1964 *The Governorship of Spanish Florida, 1700–1763.* Duke University Press, Durham, North Carolina.

Terreros y Pando, Esteban de

1788 *Diccionario castellano con las voces de ciencias y artes y sus correspondientes en las tres lenguas francesa, latina, e italiana.* Viuda de Ibarra, Madrid. Electronic document, http://ntlle.rae.es/ntlle/SrvltGUILoginNtlle, accessed October 2, 2015.

Thomas, Hugh

2003 *Rivers of Gold: The Rise of the Spanish Empire, from Columbus to Magellan.* Random House, New York.

Thornton, Peter

1978 *Seventeenth-Century Interior Decoration in England, France and Holland.* Yale University Press, New Haven, Connecticut.

Todolí, Ximo

2008 La fábrica de ceramic del Conde de Aranda de Alcora: Evolución del edificio durante el siglo XVIII. In *El azulejo, evolución ténica: Del taller a la fábrica. Actas del XI Congreso Annual de la Asociación del Ceramología celebrado en el Museo del Azulejo "Manolo Safont" de Onda, del 7 al 9 diciembre del 2006*, pp. 171–179. Fundación Museo del Azulejo "Manolo Safont," Asociación de Ceramología, Onda, Spain.

Townsend, Joseph

1791–1792 *A Journey through Spain in the Years 1786 and 1787; With Particular Attention to the Agriculture, Manufactures, Commerce, Population, Taxes, and Revenue of that Country; And Remarks in Passing through a Part of France*, Vols. 1–3. C. Dilly, London.

Trigger, Bruce G.

2006 *A History of Archaeological Thought*. 2nd ed. Cambridge University Press, Cambridge.

Unamuno, Miguel de

1902 *En torno al casticismo*. Fernando Fé, Madrid.

Vásquez, George L.

1995 Altamira, the Generation of 1898, and the Regeneration of Spain. *Mediterranean Studies* 5:85–100.

Vázquez Varela, J. M., and R. Risch

1991 Theory in Spanish Archaeology since 1960. In *Archaeological Theory in Europe: The Last Three Decades*, edited by Ian Hodder, pp. 25–51. Routledge, New York.

Voss, Barbara

2005 From *Casta* to *Californio*: Social Identity and the Archaeology of Culture Contact. *American Anthropology* 107(3):461–474.

2008a *The Archaeology of Ethnogenesis: Race and Sexuality in Colonial San Francisco*. University of California Press, Berkeley.

2008b "Poor People in Silk Shirts": Dress and Ethnogenesis in Spanish-Colonial San Francisco. *Journal of Social Archaeology* 8:404–432.

2012 Status and Ceramics in Spanish Colonial Archaeology. *Historical Archaeology* 46(2):39–54.

2014 *Losa Surtida*: Historical and Archaeological Perspectives on Imported Ceramics in Alta California. In *Ceramic Production in Early Hispanic California: Craft, Economy, and Trade on the Frontier of New Spain*, edited by Russel K. Skowronek, M. James Blackman, and Ronald L. Bishop, pp. 283–300. University Press of Florida, Gainesville.

Washburn, Dorothy K., William N. Washburn, Petia A. Shipkova

2011 The Prehistoric Drug Trade: Widespread Consumption of Cacao in Ancestral Pueblo and Hohokam Communities in the American Southwest. *Journal of Archaeological Science* 38:1634–1640.

Waterbury, Jean Parker

1983 The Castillo Years: 1688–1763. In *The Oldest City, St. Augustine: Saga of Survival*, edited by Jean Parker Waterbury, pp. 57–89. St. Augustine Historical Society, St. Augustine.

Wilkie, Laurie A.

2000 Culture Bought: Evidence of Creolization in the Consumer Goods of an Enslaved Bahamian Family. *Historical Archaeology* 34(3):10–26.

Index

Page numbers in *italics* indicate illustrations and tables.

Kathryn L. Ness is a historical archaeologist who studies how globalization impacts cultural identity and exchange, especially in early modern settings.

RIPLEY P. BULLEN SERIES

Florida Museum of Natural History

Tacachale: Essays on the Indians of Florida and Southeastern Georgia during the Historic Period, edited by Jerald T. Milanich and Samuel Proctor (1978)
Aboriginal Subsistence Technology on the Southeastern Coastal Plain during the Late Prehistoric Period, by Lewis H. Larson (1980)
Cemochechobee: Archaeology of a Mississippian Ceremonial Center on the Chattahoochee River, by Frank T. Schnell, Vernon J. Knight Jr., and Gail S. Schnell (1981)
Fort Center: An Archaeological Site in the Lake Okeechobee Basin, by William H. Sears, with contributions by Elsie O'R. Sears and Karl T. Steinen (1982)
Perspectives on Gulf Coast Prehistory, edited by Dave D. Davis (1984)
Archaeology of Aboriginal Culture Change in the Interior Southeast: Depopulation during the Early Historic Period, by Marvin T. Smith (1987)
Apalachee: The Land between the Rivers, by John H. Hann (1988)
Key Marco's Buried Treasure: Archaeology and Adventure in the Nineteenth Century, by Marion Spjut Gilliland (1989)
First Encounters: Spanish Explorations in the Caribbean and the United States, 1492–1570, edited by Jerald T. Milanich and Susan Milbrath (1989)
Missions to the Calusa, edited and translated by John H. Hann, with an introduction by William H. Marquardt (1991)
Excavations on the Franciscan Frontier: Archaeology at the Fig Springs Mission, by Brent Richards Weisman (1992)
The People Who Discovered Columbus: The Prehistory of the Bahamas, by William F. Keegan (1992)
Hernando de Soto and the Indians of Florida, by Jerald T. Milanich and Charles Hudson (1993)
Foraging and Farming in the Eastern Woodlands, edited by C. Margaret Scarry (1993)
Puerto Real: The Archaeology of a Sixteenth-Century Spanish Town in Hispaniola, edited by Kathleen Deagan (1995)
Political Structure and Change in the Prehistoric Southeastern United States, edited by John F. Scarry (1996)
Bioarchaeology of Native Americans in the Spanish Borderlands, edited by Brenda J. Baker and Lisa Kealhofer (1996)
A History of the Timucua Indians and Missions, by John H. Hann (1996)
Archaeology of the Mid-Holocene Southeast, edited by Kenneth E. Sassaman and David G. Anderson (1996)
The Indigenous People of the Caribbean, edited by Samuel M. Wilson (1997; first paperback edition, 1999)
Hernando de Soto among the Apalachee: The Archaeology of the First Winter Encampment, by Charles R. Ewen and John H. Hann (1998)
The Timucuan Chiefdoms of Spanish Florida, by John E. Worth: vol. 1, *Assimilation*; vol. 2, *Resistance and Destruction* (1998)

Ancient Earthen Enclosures of the Eastern Woodlands, edited by Robert C. Mainfort Jr. and Lynne P. Sullivan (1998)

An Environmental History of Northeast Florida, by James J. Miller (1998)

Precolumbian Architecture in Eastern North America, by William N. Morgan (1999)

Archaeology of Colonial Pensacola, edited by Judith A. Bense (1999)

Grit-Tempered: Early Women Archaeologists in the Southeastern United States, edited by Nancy Marie White, Lynne P. Sullivan, and Rochelle A. Marrinan (1999)

Coosa: The Rise and Fall of a Southeastern Mississippian Chiefdom, by Marvin T. Smith (2000)

Religion, Power, and Politics in Colonial St. Augustine, by Robert L. Kapitzke (2001)

Bioarchaeology of Spanish Florida: The Impact of Colonialism, edited by Clark Spencer Larsen (2001)

Archaeological Studies of Gender in the Southeastern United States, edited by Jane M. Eastman and Christopher B. Rodning (2001)

The Archaeology of Traditions: Agency and History Before and After Columbus, edited by Timothy R. Pauketat (2001)

Foraging, Farming, and Coastal Biocultural Adaptation in Late Prehistoric North Carolina, by Dale L. Hutchinson (2002)

Windover: Multidisciplinary Investigations of an Early Archaic Florida Cemetery, edited by Glen H. Doran (2002)

Archaeology of the Everglades, by John W. Griffin (2002)

Pioneer in Space and Time: John Mann Goggin and the Development of Florida Archaeology, by Brent Richards Weisman (2002)

Indians of Central and South Florida, 1513–1763, by John H. Hann (2003)

Presidio Santa Maria de Galve: A Struggle for Survival in Colonial Spanish Pensacola, edited by Judith A. Bense (2003)

Bioarchaeology of the Florida Gulf Coast: Adaptation, Conflict, and Change, by Dale L. Hutchinson (2004)

The Myth of Syphilis: The Natural History of Treponematosis in North America, edited by Mary Lucas Powell and Della Collins Cook (2005)

The Florida Journals of Frank Hamilton Cushing, edited by Phyllis E. Kolianos and Brent R. Weisman (2005)

The Lost Florida Manuscript of Frank Hamilton Cushing, edited by Phyllis E. Kolianos and Brent R. Weisman (2005)

The Native American World Beyond Apalachee: West Florida and the Chattahoochee Valley, by John H. Hann (2006)

Tatham Mound and the Bioarchaeology of European Contact: Disease and Depopulation in Central Gulf Coast Florida, by Dale L. Hutchinson (2006)

Taino Indian Myth and Practice: The Arrival of the Stranger King, by William F. Keegan (2007)

An Archaeology of Black Markets: Local Ceramics and Economies in Eighteenth-Century Jamaica, by Mark W. Hauser (2008; first paperback edition, 2013)

Mississippian Mortuary Practices: Beyond Hierarchy and the Representationist

Perspective, edited by Lynne P. Sullivan and Robert C. Mainfort Jr. (2010; first paperback edition, 2012)
Bioarchaeology of Ethnogenesis in the Colonial Southeast, by Christopher M. Stojanowski (2010; first paperback edition, 2013)
French Colonial Archaeology in the Southeast and Caribbean, edited by Kenneth G. Kelly and Meredith D. Hardy (2011; first paperback edition, 2015)
Late Prehistoric Florida: Archaeology at the Edge of the Mississippian World, edited by Keith Ashley and Nancy Marie White (2012; first paperback edition, 2015)
Early and Middle Woodland Landscapes of the Southeast, edited by Alice P. Wright and Edward R. Henry (2013)
Trends and Traditions in Southeastern Zooarchaeology, edited by Tanya M. Peres (2014)
New Histories of Pre-Columbian Florida, edited by Neill J. Wallis and Asa R. Randall (2014; first paperback edition, 2016)
Discovering Florida: First-Contact Narratives from Spanish Expeditions along the Lower Gulf Coast, edited and translated by John E. Worth (2014; first paperback edition, 2015)
Constructing Histories: Archaic Freshwater Shell Mounds and Social Landscapes of the St. Johns River, Florida, by Asa R. Randall (2015)
Archaeology of Early Colonial Interaction at El Chorro de Maíta, Cuba, by Roberto Valcárcel Rojas (2016)
Fort San Juan and the Limits of Empire: Colonialism and Household Practice at the Berry Site, edited by Robin A. Beck, Christopher B. Rodning, and David G. Moore (2016)
Rethinking Moundville and Its Hinterland, edited by Vincas P. Steponaitis and C. Margaret Scarry (2016)
Handbook of Ceramic Animal Symbols in the Ancient Lesser Antilles, by Lawrence Waldron (2016)
Paleoindian Societies of the Coastal Southeast, by James S. Dunbar (2016)
Gathering at Silver Glen: Community and History in Late Archaic Florida, by Zackary I. Gilmore (2016)
Cuban Archaeology in the Caribbean, edited by Ivan Roksandic (2016)
Archaeologies of Slavery and Freedom in the Caribbean: Exploring the Spaces in Between, edited by Lynsey Bates, John M. Chenoweth, and James A. Delle (2016)
Setting the Table: Ceramics, Dining, and Cultural Exchange in Andalucía and La Florida, by Kathryn L. Ness (2017)

www.ingramcontent.com/pod-product-compliance
Lightning Source LLC
LaVergne TN
LVHW011815200926
848484LV00005B/1209
* 9 7 8 1 6 8 3 4 0 0 0 4 2 *